Susanne Saygin

'AN IDEA IS ONLY GOOD IF YOU TURN IT INTO REALITY'

FRITZ MÜLLER – THE BIOGRAPHY OF AN ENTREPRENEUR

THORBECKE

www.thorbecke.de • info@thorbecke.de

Concept and editing: Geschichtsbüro Reder, Roeseling & Prüfer, Cologne
Translation: Aingeal Flanagan, Flanagan Language Services, Cologne
Design: Jefferson & Högerle, Studio for Design, Cologne
Concept: Hans Schlimbach, Cologne

Printed in Germany

ISBN 978-3-7995-0224-5 (german edition)

Susanne Saygin

'AN IDEA IS ONLY GOOD IF YOU TURN IT INTO REALITY'

FRITZ MÜLLER – THE BIOGRAPHY OF AN ENTREPRENEUR

THORBECKE

CONTENTS

INTRODUCTION	The courage to be independent	06
CHAPTER 1	Prologue: A family of entrepreneurs	08
EXCURSUS	'We are kings and princes here' The Hohenlohe region and the people who shaped it	18
CHAPTER 2	Widowed with children: Marie Müller in Ingelfingen (1942–53)	30
CHAPTER 3	Making something out of nothing: Fritz Müller's childhood and youth in the Kocher Valley (1942–53)	42
CHAPTER 4	Apprenticeship and early professional experience (1953–63)	54
CHAPTER 5	The idea: a valve made of plastic	74
CHAPTER 6	Home-made valves: the early years (1963–65)	88
CHAPTER 7	Early successes: GEMÜ's growth and development (1965–68)	100

CHAPTER 8	Continuous growth: GEMÜ in the 1970s	116
CHAPTER 9	'New products for new markets': GEMÜ's international expansion (1981–90)	132
EXCURSUS	The promotion of vocational training	148
CHAPTER 10	Expansion instead of standstill: high Purity, expansion into Asia, and the runaway success of aseptic products (1991–98)	158
EXCURSUS	Having strong local roots in a globalized world	178
CHAPTER 11	New beginnings: international challenges and the GEMÜ Dome (1998–2008)	192
INTERVIEW	'An idea is only good if you turn it into reality'	210
NOTES	Acknowledgements	220
	Selected bibliography	222
	Picture credits	228

INTRODUCTION

THE COURAGE TO BE INDEPENDENT

Growing up in the rural Kocher Valley in the early 1950s, Fritz Müller was thrilled by two things in particular as a ten to twelve year old boy: the gracefulness of his self-made model airplane as it glided over the vineyards of his home town, Ingelfingen, and the speed of his self-made cart as it carried him to victories in soap-box derbies. Looking back, it now seems as if the speed of his soap-box cart and the gracefulness of his airplane meant two things for the young boy: independence and freedom. These things in turn shaped his ideas of what he wanted to achieve in his lifetime: a racing car, an airplane, and a company of his own. Today, almost 60 years on, Fritz Müller owns a number of Ferraris, numerous vintage cars, and several racing cars; he also owns three airplanes, one of which he uses to fly to the foreign subsidiaries of his company, GEMÜ, a global operation that manufactures valves and the measurement and control systems needed to operate them.

This book traces the various stages along the road Fritz Müller travelled in order to turn his ideas into reality. The purpose is not to celebrate his status as a high-flyer who realised his dreams by going it alone. While it is true that Fritz Müller single-mindedly went his own way, he neither started the race from pole position, nor did he lead the field from start to finish without ever being challenged along the way. This biography records the hardships of Fritz Müller's childhood as the son of a war-widow in the Kocher Valley in post-war Germany and documents the difficulties and the many ups and downs he has experienced throughout the course of his working life. Chapter for chapter, this book shows how Fritz Müller realised his dream of establishing his own company. It is Fritz Müller's hope that this book will encourage young people in the early stages of self-employment to continue along their chosen path. It is intended to be both a practical manual for what is an undoubtedly difficult course of action and a source of moral support and inspiration in trying times.

This biography is not aimed exclusively at entrepreneurs. It is also intended to encourage young people starting out in life to actively implement their ideas and follow their dreams without being discouraged by difficult situations or hampered by exaggerated expectations of success. The message behind this book is that all those who take the initiative and assume responsibility for their actions are independent and are, to a certain extent, entrepreneurs. A large part of GEMÜ's success is also due to the fact that down through the years, Fritz Müller received support from a number of people who contributed their ideas, talents, and energy to the company. This book bears witness to Fritz Müller's appreciation of what these people did and is intended to be an inspiration to the next generation.

And so, Fritz Müller is doing for the next generation what his grandfather – himself a successful entrepreneur – did for him almost 70 years ago when he took every opportunity to encourage his grandson with the words: 'Fritz, see to it that you set yourself up in business.' Fritz Müller's desire to pass on this message to young people can be seen in the entrepreneur's manifold activities and it is hoped that through this book, this message will reach an even greater number of young people. By picking up where his grandfather left off and by sharing the wealth of experience he has gained down through the years with those outside his immediate circle, he would like to continue a tradition started by other entrepreneurs whose activities shaped the history of the Hohenlohe region down through the centuries.

CHAPTER 1

PROLOGUE: A FAMILY OF ENTREPRENEURS

Fritz Müller Jr.'s paternal grandfather, the master plasterer Friedrich Müller

It was the same thing every time he went to see his grandfather in Esslingen. Even though Fritz Müller was still at school and not yet of college age, Friedrich Müller would turn to his grandson and say: 'Boy, see to it that you set yourself up in business. It doesn't matter if it's only a small workshop, just set yourself up in business.' Friedrich Müller knew what he was talking about. Born the son of a simple shoemaker sometime around 1880 in Denkendorf in what is now the state of Baden-Württemberg, he learned the plastering trade, obtained a master craftsman's diploma, and set himself up in business in the nearby city of Esslingen sometime around 1900. In the early days he had little more than a wheelbarrow, a bundle of scaffolding poles, and a few buckets, trowels, and shovels. But the young plasterer, who specialised in outdoor plasterwork, was ambitious and hard working and made the most of the many opportunities in the building trade which flourished during the reign of William II. Within the space of only a few years, Friedrich Müller had transformed his one-man operation into a thriving company. By 1931 he had more than 50 employees and was accepting orders from customers far beyond Stuttgart, the next major city to Esslingen. Friedrich Müller liked to keep his plasterers on their toes by booming out instructions like: 'C'mon! Get a move on up there!', but his brusque manner belied the fact that he was a kind and fair employer who knew every one of his employees personally and was always ready to lend an ear. This earned him the unswerving loyalty of his workers. 'If ever anyone had a bad word to say about grandfather,' recalls Fritz Müller, 'they were in for trouble with our 52 plasterers. They stood steadfastly behind their boss.'

Friedrich Müller applied the same blend of strictness and fairness to his dealings with his employees as he did to the raising of his children. As far as he was concerned, there was never any doubt that his son Fritz (*b.* 1904), would one day take over the company. Nevertheless, he did not stand in the way when Fritz decided to become a mechanic. It was only after he had qualified as a mechanic that Fritz Müller became an apprentice plasterer in his father's business. In 1921, before he had even turned 18, he obtained his master craftsman's diploma in plastering.

It seemed as if Fritz Muller's professional life was all mapped out in front of him: he would join his father's thriving company and one day take over at the helm. Many others would have been delighted at such a prospect, but not Fritz Müller. He had his sights set on enhancing his qualifications and becoming a stucco plasterer, which would attract lucrative jobs involving the intricate interior decoration of bourgeois villas. His father did not share his vision. Yes, he agreed, it would be very nice to be able to offer customers stucco plasterwork, but the order books were already full as it was. And anyway, there was no time for further training; his son was needed in the business right away.

But Fritz Müller had other ideas and was not about to be dictated to by his father. His son, Fritz Müller Jr., relates what happened: 'One day, Aunt Maria went to wake my father. She went to the bottom of the stairs and shouted up: "Fritz, come down, the coffee is ready!" But there was no sign of Fritz. So she went upstairs to see what the matter was and found his bedroom empty. My father had upped sticks in the middle of the night and left with a friend, leaving nothing – absolutely nothing – behind. He was gone! Just like that! My grandfather, rough diamond that he was, was beside himself with rage: "Don't let him set foot in this house again! I'll cut him off without a penny! I never want to set eyes on him again," and so on and so forth.'

Friedrich Müller remained intransigent even when his son wrote a few weeks later to say that he had started an apprenticeship as a stucco plasterer in a town near Lake Constance. It is likely, however, that he respected his son's determination despite himself. The fact that Fritz had not allowed himself to be diverted from his course and had the courage of his convictions despite vehement opposition, showed independence of character and an ability to assert himself. His father knew that these two characteristics were vital for anyone running a large business such as theirs. When Fritz Müller returned to Esslingen two years later with his master craftsman's diploma in stucco plastering, and decorated his parents' house with stucco work, his father could not hide his pride. 'That's my boy!' he would say. His boy had proven that he could stand on his own two feet.

Fritz Müller on a motorcycle excursion with friends. The BMW R 11 with sidecar was used during the week to transport material. To the dismay of his father, Friedrich, Fritz Müller used the company motorcycle not only for leisurely day-trips, but also for motorcycle rallies. 'All hell broke loose on Monday mornings when the motorbike was found to be in need of repair,' says Fritz Müller Jr.

A short time later, Fritz Müller finally joined his father's company. There was certainly no shortage of work: the order books were full, not least because the Weimar Republic ran a number of public building programmes. What is more, Friedrich Müller had significantly increased the area in which his company worked. In the 1920s, his plasterers were working

Marie Hasel in the side car of the Müller company motorcycle; Fritz Müller, her future husband, stands behind her (c.1930)

in places as far away as the Kocher Valley, which was almost 100 km away from Esslingen. Here, under the watchful eye of Fritz Müller, they plastered the electricity company's new buildings along the river Kocher and the new railway stations on the final stretch of the Kocher Valley Railway line, construction of which had begun in 1892. The materials needed for such large jobs as these – scaffolding, tubs for mixing mortar, tools, sand, and lime – were no longer transported in wheelbarrows, but delivered by rail in open wagons to the nearby town of Künzelsau.

Fritz Müller was obviously very taken with the picturesque Kocher Valley and its many vineyards because he returned there at the weekends on his motorcycle with friends even after the company had finished working in the region. It is likely that he met Marie Hasel of Ingelfingen, who was the same age as himself, on one of these motorcycle excursions.

Marie Hasel was the daughter of the cellar master of the Ingelfingen Wine-Growers' Co-operative. Due to its excellent location in the Kocher Valley, Ingelfingen has a centuries-long tradition of wine-making. In fact, for a long time wine-making was the chief source of income for the people of the town. However, aging vines and errors of cultivation and blending in the nineteenth century led to a drop in both the yield and quality of Ingelfingen's wine. The situation was further aggravated by a lack of storage capacity that repeatedly forced Ingelfingen's wine-growers to produce their wines too quickly and, consequently, sell them at a low price.

By the early 1890s, wine-growing in Ingelfingen was in decline. It was agreed that the only way out of the predicament was for the wine-growers of the locality to join forces and make some essential changes. On the initiative of the mayor of Ingelfingen, Joseph Rilling, and under his leadership, approximately 30 wine-growers in Ingelfingen founded the Wine-Growers' Co-operative in September 1892. In the years that followed, this co-operative completely transformed viticulture in the locality. With the help of grants from the state, the co-operative purchased new wine-presses, increased storage capacity, began to improve methods of cultivation, and hired a professional cellar master.

Fritz Müller's grandfather, Christian Hasel (1873–1934), was appointed to this important post in 1894. Hasel hailed from one of a number of Ingelfingen's small farming families, whose main source of income was wine-growing, but who also cultivated potatoes and cereals on a small scale. Small as their farm might have been, the family was obviously wealthy enough to pay for Christian to train as a cooper and to subsequently attend the School of Viticulture in Weinsberg (now the State Education and Research Institute for Viticulture and Pomiculture).

With such a comprehensive education as this, Hasel had exactly the right wide-ranging knowledge that was required of a cellar master in the

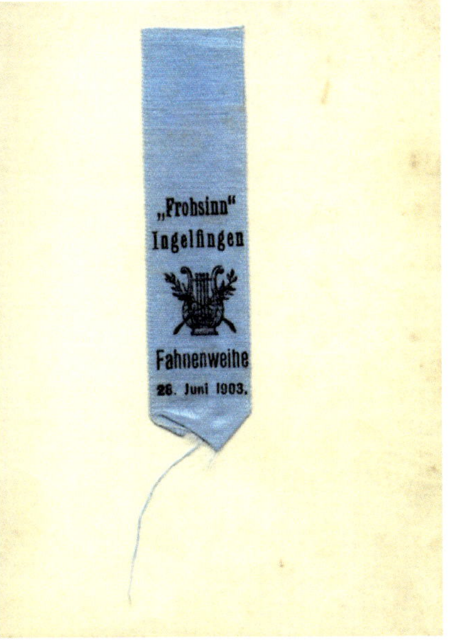

One of Christian Hasel's note-books, which dates from the turn of the century and has been handed down from generation to generation in the family, containing poems (probably written by Hasel himself), drawings of buildings and places in the regions, as well as notes on choral festivals in the Kocher Valley, shows that Christian Hasel was interested in a variety of cultural activities and was firmly rooted in the life of the community of Ingelfingen

Ingelfingen Wine-Growers' Co-operative. The cellar master was responsible for checking the quality of the harvested grapes, pressing the grapes, managing and checking the fermentation, and the blending and maturing of the wine in barrels. This meant that Christian Hasel's job was not only vitally important for the wine-growing sector in Ingelfingen, it was also a very thankless one. 'If the wine turned out well, the praise went to the wine-grower; if it turned out poorly, the cellar master was to blame,' says Fritz Müller Jr. Much was expected of the cellar master, especially when the harvest was poor and the wine-growers began to push for an abandonment of the co-operative's quality standards so that they could absorb at least some of their losses.

In the first years of its existence, the Ingelfingen Wine-Growers' Co-operative had to deal with the animosity of those wine-growers who had decided not to join the co-operative. A report from the year 1895 shows just how high emotions could run. The first harvest during Christian Hasel's time as cellar master was so good that the Wine Growers' Co-operative decided to submit the wine for the DLG wine awards in Stuttgart. An Ingelfingen wine-grower who had chosen not to join the co-operative publicly claimed that the wine submitted to the jury in Stuttgart had not been produced by the co-operative at all, but had secretly been bought-in. He claimed that the wine produced by the local co-operative was in fact cloudy. Despite these scandalous claims, the co-operative was able to acquit itself and successfully defend its reputation. The wine-grower had to publicly retract his accusation of fraud and pay a fine.

In a town the size of Ingelfingen, where everyone knew everyone else, the cellar master not only needed to keep an eye on the interests of the co-operative and to be able to assert himself in order to defend these interests, he also needed to understand the concerns and the economic situation of small wine-growers. The fact that Christian Hasel held the post of cellar master at the Ingelfingen co-operative for just under 40 years (from 1894 to 1931) and was nominated to the board of the co-operative in the final years of his life, is evidence of the fact that he was both able to assert himself and had a sense of responsibility. He was also very successful: the quality of the wine improved and, slowly but surely, the wine-growers of Ingelfingen began to fetch better prices for their wines.

Fritz Müller came to Ingelfingen every weekend to court Marie Hasel. It is possible that she saw in her beau the entrepreneurial qualities her father possessed. Whether this was indeed the case or not, she married the young master plasterer and stucco plasterer in June 1931 and moved with him to Esslingen. Their first son, Kurt, was born in 1935; their second son, Fritz Dieter, was born on 2 January 1939.

The young father and entrepreneur had ambitious plans for expanding his father's business. During both his apprenticeships, he had become familiar with the various aspects of the building trade and now dreamed of combining the various services associated with building to provide

Cellar master Christian Hasel in the wine cellar of Ingelfingen's Wine-Growers' Co-operative

A wine auction organised by Ingelfingen's Wine-Growers' Co-operative in front of the new castle: Fritz Müller Jr.'s grandfather, cellar master Christian Hasel (second from left), confidently observes the buyers

Fritz Müller enjoying his last leave of absence with his wife Marie and children Fritz (left) and Kurt (right)

customers with a complete package. In order to realise his idea of delivering turnkey houses from a single source, he would have to employ not only plasterers, but also bricklayers, painters, electricians, and fitters.

The ink was barely dry on the employment contract for the company's first new employee when the German army marched into Poland in September 1939, heralding the start of the Second World War. Business for the Müllers boomed like never before. Countless new barracks had to be plastered and finished. It was a case of all hands on deck. The junior boss, Fritz Müller , was particularly busy. It was for this reason that Friedrich Müller arranged for his son to be exempted from military service. Fritz was furious when he found out what his father had done. Like many Germans at the start of the war, he was convinced that German troops would secure a swift victory. 'We're going to win the war anyway, and I don't want to be called a conchie when it's all over,' he is reported to have said to his father before volunteering for the mountain infantry as a medical orderly.

Fritz Müller returned home on leave of absence for the last time in 1940. He had already experienced enough to know that his initial bold optimism had been misplaced. But it was too late; the young father had no choice but to return to the front. In 1941, Fritz Müller, aged 36 years, was killed in action in the Balkans while attempting to rescue an injured comrade.

A FEW WORDS ON ...

'WE ARE KINGS AND PRINCES HERE' THE HOHENLOHE REGION AND THE PEOPLE WHO SHAPED IT

We are kings and princes here' is how Fritz Müller describes the social and economic role played by entrepreneurs in the Hohenlohe region. What may seem boastful or even conceited at first is in fact a surprisingly appropriate comparison, especially when one considers the history of the region and above all the social and economic responsibility that has always been borne by those personalities who shaped the Hohenlohe region down through the centuries. From the Enlightenment to their enforced abdication in the early nineteenth century, the princes of Hohenlohe encouraged the settlement of new sectors of trade and industry in the region, thereby promoting its social and economic development. This role was adopted in the late nineteenth century by civil politicians, who became the pacemakers of modernisation. Since the end of the Second World War at the latest, it has been the region's companies – many of which are global, family-run companies such as Fritz Müller's GEMÜ – that have created jobs and apprenticeships and have through their active social involvement in the region helped turn Hohenlohe, which was ridiculed as 'Swabian Siberia' right up until the 1970s, into a booming growth region.

The cultural area that is Hohenlohe – which is situated between the Bauland, Odenwald, schwäbisch-fränkischer Waldberg and Frankenberg regions – has been an area of stark contrasts since time immemorial. In ancient times, the limes running from Jagsthausen via Öhringen to Main-

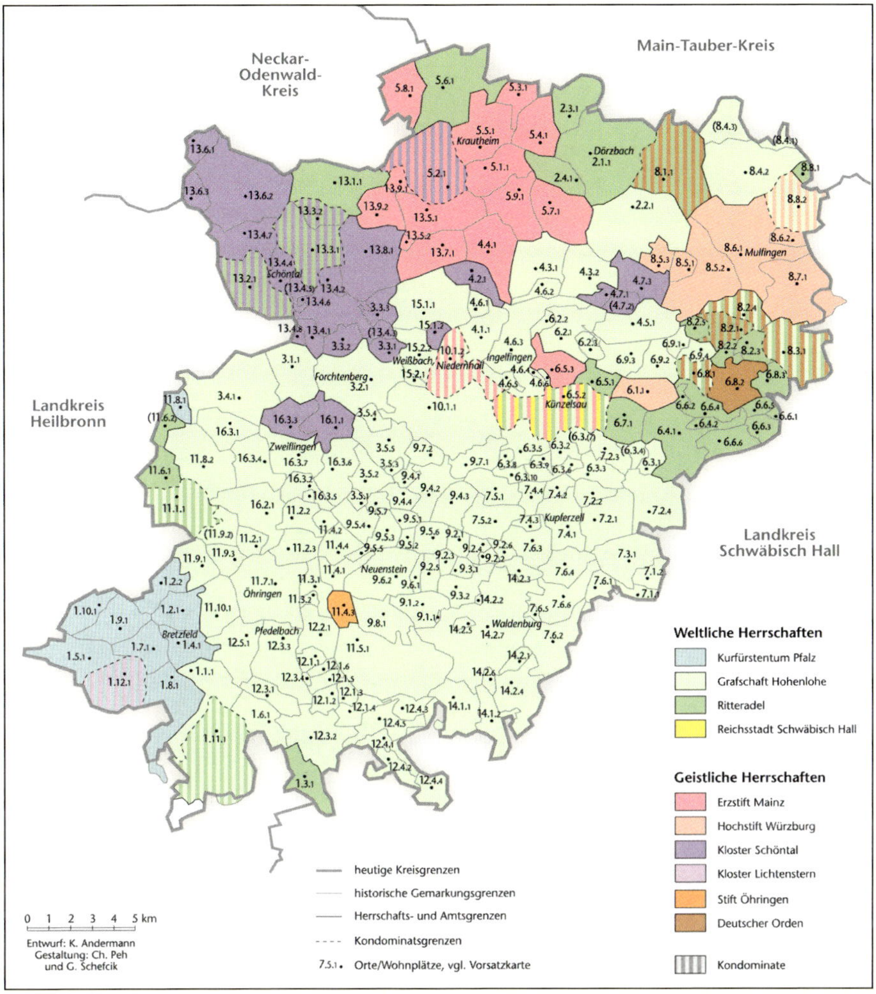

hardt marked the boundary between the Roman Empire and unoccupied Germania. Centuries later, the settlements of the Alemanni and the Franks overlapped in the area between the rivers Neckar, Tauber, and Kocher. To this day, the river Kocher and the river Jagst form the linguistic border between the Alemanni-Swabian and the South- and East-Franconian dialects.

The earldom of Hohenlohe and the surrounding regions c.1500: localities such as Niedernhall and Künzelsau were governed by a number of lords simultaneously

The Hohenlohe family tree

In the Middle Ages, governance of this sparsely populated stretch of land was shared by a number of Church and secular rulers. The House of Hohenlohe – which derived its name from the family seat, a fortified castle near Uffenheim known as 'Hohlach' or 'Hohenloch', which no longer exists – is first mentioned in deeds dating from the twelfth century. Sometime around the year 1330, the House of Hohenlohe ruled over the region between Heilbronn in the West, Schillingfürst in the East, the Waldenburger mountains near what is now the city of Schwäbisch Hall in the South, and the river Jagst in the North. This area of influence did not form a political or economic unit, but comprised a number of different regions and systems of law. The Hohenlohes extended this sphere of influence through marriage. However, it was also through marriage, namely in the form of the payment of dowries for the daughters of the family, that it lost territory too. Pledges and the sale of rights and estates further eroded the power of the House of Hohenlohe. This process of erosion could not be stopped by either the establishment of entailed estates, which afforded the right of use to only one member of each of the noble families that were given earldoms in the late fifteenth century, or the so-called 'division of land' in 1553, which divided up the Hohenlohe sphere of influence between the two main family lines in the house, the Waldenburgs and the Neuensteins. The earls of Neuenstein and Waldenburg continued dividing up their respective property down through the generations, which meant that in the centuries that followed, up to twelve lines of the same family ruled in the region simultaneously. The gulf that already separated the two main families was deepened by religious differences, the Neuensteins being supporters of the Protestant faith and the Waldenburgs loyal to Catholic Rome.

The continuing religious and regional fragmentation led to tension and disputes over inheritance in some areas and to economic progress in others. In view of the fact that each of the tiny Hohenlohe 'kingdoms' wanted to demonstrate its importance to the world, representative castles and fortified houses were built throughout the region. These buildings

Die Linien und Zweige des Hauses Hohenlohe seit dem Ende des Mittelalters

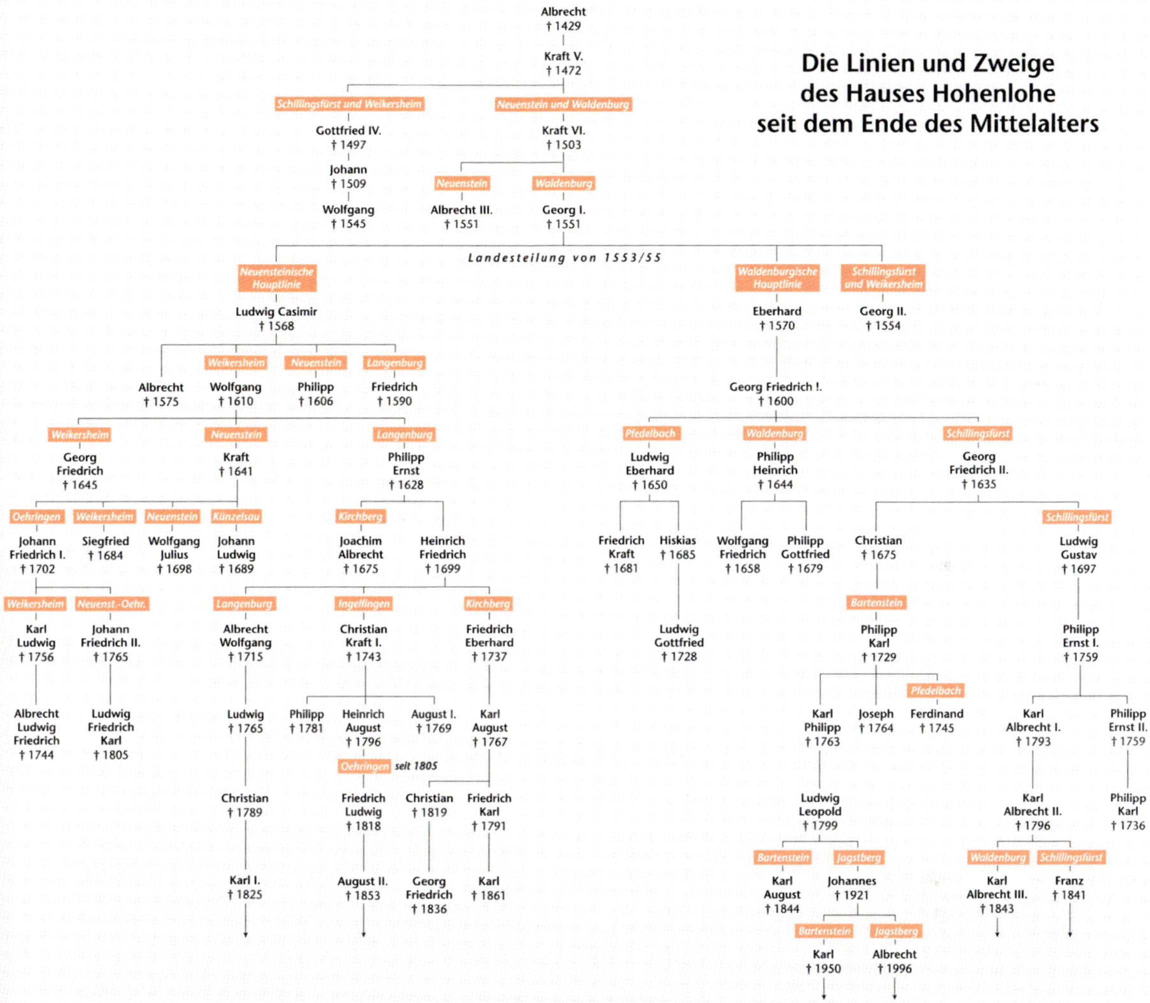

mean that the region is now a wonderful destination for tourists seeking historical sites. Overall, however, this fragmentation had a negative effect on the region because none of the cities which served as the seat of a local noble family could really develop into a centre of any kind of power. Öhringen remained the most important town in the region well into the eighteenth century with its 4,000 inhabitants; Waldenburg, Weikersheim, or Ingelfingen, on the other hand, were much less populous. Most of the Hohenlohe's 100,000 subjects lived in hamlets and small villages.

The development of the region was above all hindered by its poor transport links, which remain a problem for the local economy to this day. With the exception of the old trade and post highway from Nuremberg to Heilbronn, which passed through Öhringen, there were no roads worth speaking of. Many of the region's smaller localities were only connected to each other by unbound roads until well into the eighteenth century.

'A castle in every village': the numerous representative castles and fortified houses in the region, such as the one pictured here in Neuenstein, are evidence of the self-confidence of the Hohenlohe rulers

Traders and merchants from other parts of the country avoided the region between the rivers Neckar, Tauber, and Kocher; even the locals were reluctant to travel.

The Hohenlohe rulers left the administration of their scattered estates to their stewards and bailiffs. In theory, these men had far-reaching powers; in practice, however, they did not have instruments of power that were effective enough to allow them to assert their influence over the farmers, trade workers, or citizen farmers in the locality. In 1525, for example, the Hohenlohe's subjects showed how violently they could react to demands they considered inappropriate. When the earls imposed heavy taxes despite the fact that the region had suffered poor harvests, thousands of farmers in the region between the rivers Neckar, Tauber, and Jagst rose up against their lords under the leadership of the Hohenlohe's chief mercenary, Götz von Berlichingen, and plundered monasteries and the homes of the nobility. It took the lords several weeks to put together a hastily convened army of mercenaries to crush the rebellion. Thousands of farmers were slaughtered; anyone who survived was crippled with heavy fines.

Land of inventors: when the Hohenlohe's chief mercenary, Götz von Berlichingen, lost his right hand in battle, he had a metal prosthesis made; the complex engineering that went into this 'iron hand' allowed von Berlichingen to grasp objects, hold them, and move his arm

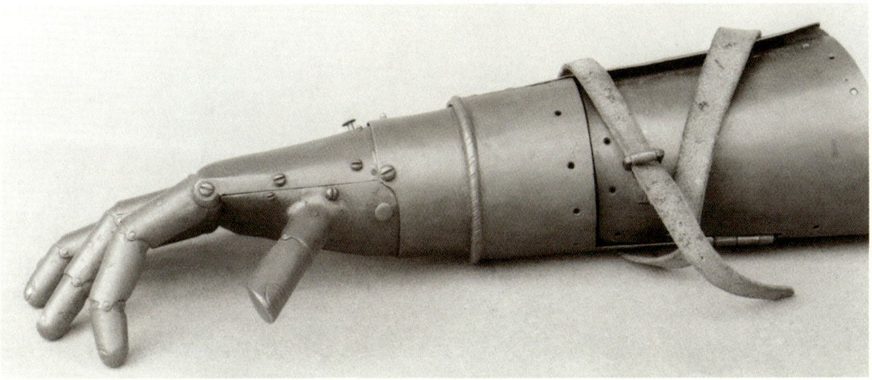

Despite the fact that the farmers' rebellion had been brutally put down, it remained hard for the Hohenlohe's stewards and bailiffs to push

Frederick Louis, Prince of Hohenlohe-Ingelfingen, and his wife Marianne von Hoym zu Droysig wanted to release Ingelfingen from its dependence on wine-growing by encouraging trade and businesses to settle in the locality

through the interests of their lords. Because crop farming and the breeding of livestock were almost impossible on the steep slopes of the Jagst, Tauber, and Kocher valleys, most people in the region earned their living by growing grapes and making wine. However, the early frosts of autumn and the poor transport links made it very difficult for them to sell their wines. Moreover, just like their lords, the wine-growers in the Hohenlohe region divided their land among their offspring which meant that as time went on, the plots of land held by each family became smaller and smaller, generating barely enough income to support the families that lived off them. It was, therefore, very difficult to collect taxes from these people. Over the centuries, resistance to 'the lords of the manor' hardened. Even the well-off farmers who worked the fertile land of the Hohenlohe plateau were confident enough to refuse to pay their taxes or do the unpaid labour demanded by their lords because they knew the latter could not force them to do otherwise. With subjects like these, it was very difficult to build up a modern state. Nevertheless, some local lords, such as the ambitious reformer Frederick Louis, Prince of Hohenlohe-

Ingelfingen (1746–1818) who had learned the ideals of the enlightenment from Frederick II of Prussia, sought to promote the local economy in the Prussian style.

In order to attract businesses to his town of residence, Ingelfingen, Frederick Louis reserved a piece of land outside the town walls for what could be referred to as a pre-industrial business park. He named the site Mariannenvorstadt in honour of his wife and sought tenants by offering 'free space to build, no charges, freedom for ten years, no unpaid labour, no slavery, no soldiers'. The prince also praised the pleasant location of the site and the mild climate of the town on the river Kocher. Initially, very few showed interest in the site. After a few years, however, the influx of businesses was so great that only the best applicants were assured a plot of land in Mariannenvorstadt. Inscriptions on houses dating from this period such as 'Long live his Highness Prince FL' or 'FL. Your name shall last forever' bear witness to the gratitude of the watchmakers, brewers, paper-makers, hosiery workers, nail-makers, and gunsmiths who set up shop in Mariannenvorstadt.

The pre-industrial business park set up in Mariannen- vorstadt in the 1780s was the first of its kind in Ingelfingen

The inhabitants of Ingelfingen also benefited from their lord's enthusiasm for reform. One of the prince's advisors remarked: 'Foreign settlers are rarely as good [...] as the natives,' and suggested that Ingelfingen's 'boys should be taught such trades and professions as befit their inclinations and abilities and would benefit the locality by their duplication.' The education and training of these boys could, he suggested, be financed by foundations. He went on to say that talented youths from the town should receive their training elsewhere and return to their native town to pass on the knowledge they had gained to the next generation. The prince also made efforts to improve the school system. In the 1790s, some 300 boys and girls attended the school at Church Gate in Ingelfingen. Even in the late eighteenth century, the importance of highly qualified workers for the economy was clearly recognised and promoted.

Frederick Louis's reforms seemed to be taking effect. Ingelfingen was experiencing an unprecedented cultural and economic upswing. This upswing was, however, completely dependent on the support of the local prince. When Frederick Louis inherited the principality of Hohenlohe-Öhringen in the summer of 1805 and moved his residence to the much larger town of Öhringen, the people of Ingelfingen beseeched their prince to 'graciously heed the pain that the loss of his residence would mean for this town'. Their appeal fell on deaf ears. Even if it had not, however, Ingelfingen would

The document that recorded the honourable discharge of Fritz Müller's ancestor Kristian Hasel from the Württemberg army in 1820

not have remained the prince's residence for much longer. Following the defeat of Prussian troops near Jena and Auerstedt in October 1806, Napoleon banished Frederick Louis, Prince of Hohenlohe-Ingelfingen, who had been the senior commander in the Prussian army, from the region. The principality of Hohenlohe was dissolved in 1806 and integrated into the Kingdom of Württemberg, which had previously allied itself with Napoleon.

The new lords of the region considered Hohenlohe to be something of a bread-basket and a reservoir of troops. Before the end of 1807, the young men of Hohenlohe were pressed into military service in the Württemberg army. Kristian Hasel, an ancestor of Fritz Müller, was one of these young men. Drafted into the army in March 1813 just before his seventeenth birthday, Hasel fought as a soldier against Napoleon in the War of the Sixth Coalition between 1813 and 1815 (the King of Württemberg changed sides shortly before the Battle of Leipzig in 1813) and was discharged in the spring of 1820.

During his absence, his native town of Ingelfingen had rapidly gone into decline. Many tradesmen left Mariannenvorstadt causing a slump in housing prices, and the economy shrunk back to the level it was at before Frederick Louis introduced his reforms. The last great famines to hit continental Europe in the years 1816/17 and 1846/47 dealt a severe blow to Ingelfingen. Because the King of Württemberg did little to relieve the suffering in the region, the people of Hohenlohe rose up in rebellion against their sovereign in March 1848. However, the enthusiasm for rebellion ended after a brief occupation of the bursary and a number of political meetings. In 1849, a period of conservative peace set in not only in Ingelfingen but across all German states. Despite the peace, the situation of the people had not improved, and so, the wave of emigration continued and grew in the years after 1848. Between 1806 and 1870, almost a third of the population of Ingelfingen (almost 500 people) left the town. Those who remained, struggled with economic stagnation.

One Württemberg bailiff wrote: 'greater commercial enterprise, factory production, and the exploitation of mineral resources, which has, as yet,

Lost glory: when the prince moved his residence to Öhringen in the early nineteenth century, Ingelfingen slid into a period of economic stagnation that would last almost a full century

failed to materialise, could raise the fortunes of this little town.' Ultimately, it was a man who was not born in Ingelfingen who released the town from its resignation. Joseph Rilling originally came to Ingelfingen as a clerk of the court. Elected mayor of the town in 1892, he spent his 27 years in office radically modernising Ingelfingen. In addition to lending his support to the wine-growers' association, Rilling saw to the establishment of a trade association (1898), the Dairy Co-operative (1899), and the Pomiculture Association (1910). As members of the credit union founded by Rilling in 1893, locals could invest any money they had saved and take out low-interest loans. Rilling became known as the 'beggar of Ingelfingen' for his talent at extracting funding from public authorities for his projects, one of which included the setting up of a new secondary school in the town. He also put his weight behind the development of the spa and cure sector and attracted new industries to the locality. Rilling was also

Elected mayor in 1892, Joseph Rilling spent almost 30 years energetically modernising the town of Ingelfingen

responsible for other important projects such as the construction of a sewerage system, the maintenance of local roads, and the linking up of Ingelfingen to the Kocher Valley railway line. Work on the railway line was, however, delayed by the First World War and the period of inflation that followed it, and the station was only opened for traffic in 1924.

In less than 30 years, what had once been a backward little town had caught up with the twentieth century. In 1919, Rilling was succeeded as mayor of Ingelfingen by his son Eugen, who continued his father's economic and political policies throughout the difficult years of the Weimar Republic. A vine improvement association was set up, the town was at last linked up to the railway network, and the spa and cure sector began to take off. An electric power station was built between 1919 and 1922 in order to supply the Kocher Valley with electricity. A canning factory was also opened in the town around the same time. Other small businesses followed around 1930. It is possible that Eugen Rilling was an even more popular mayor than his father had been before him; in 1929 he was voted back into office when he received 597 of a total 618 votes. He remained in office until he was nominated mayor of the district capital, Welzheim.

However, without the people of Ingelfingen themselves, the advances made during the 'Rilling era' would never have been possible. They learned that the future does not lie in rigidly sticking to outdated structures, but in pulling together to implement new ideas. This was the foundation for the economic success of companies like those run by the Reisser brothers, Christian Bürkert, and Fritz Müller, who all established companies in Ingelfingen after the Second World War and who, not least through their promotion of education and professional training and their active social involvement, have helped keep Frederick Louis's enlightened ideals alive in Ingelfingen to this day.

CHAPTER 2

WIDOWED WITH CHILDREN: MARIE MÜLLER IN INGELFINGEN (1942–53)

When her husband was killed in action, Marie Müller had to raise her two sons alone

The death of her husband abruptly turned Marie Müller's life on its head. Up until then, her position in the Müller family company and that of her children had been clearly defined: she was the wife of the junior manager, who would take over his father's business in the foreseeable future. Fritz Müller had already proven that he had what it took to manage the company successfully after his father's retirement. Moreover, it was to be expected that at least one of her sons would lead the family company into the third generation.

These prospects faded with the death of her husband. In 1941, her sons, Kurt and Fritz-Dieter, were six and two years old respectively. This meant that it would be years before one of them could join the family company and support their grandfather. But until that day, how was Marie Müller to support herself and her sons? The modest pension to which she was entitled as a war-widow was barely enough to buy essentials. Her father-in-law Friedrich Müller would have supported her and her sons, but Marie Müller did not want to have to rely on his assistance. What's more, she was born the daughter of a wine-grower in Ingelfingen, a rural community, and had never really settled into urban

Marie Müller with her two sons and one of her nieces in Ingelfingen (c.1943)

life in Esslingen. The death of her husband obviously increased her desire to return to her native town.

What worried Marie Müller the most about living in Esslingen was the safety of her children. There had been isolated air raids on Stuttgart as early as 1940, and the number of raids began to increase in the autumn of 1941. However, it was in May 1942 that an attack led by a major British bomber squadron caused widespread death and injury for the first time. It was only to be expected that Stuttgart, an industrial city, would remain the focus of allied bombing raids. For this reason, in the

The widow Marie Müller (left) with her sons, her brother, and his family in Ingelfingen; her brother and sister-in-law did what they could to support her

spring of 1942, Marie Müller and her sons moved back to Ingelfingen, where she initially lived with her brother, Christian Hasel, and his family.

Her relatives did all they could to help her build a new life for herself and her children. Marie Müller was in her late thirties when she returned to Ingelfingen. Although she had gone into service in houses in the Black Forest as a young girl, like many women of her generation, she had never actually learned a profession, but had stayed at home to look after the children. Nevertheless, she was willing and able to work. There was plenty of work in Ingelfingen. Following a series of poor harvests in the 1920s, the wine-growers had been blessed with a number of at least mediocre – and some good – harvests in the late 1930s. Moreover, a number of smaller companies had been set up in the town: a timber goods factory, a textile manufacturer, and a small

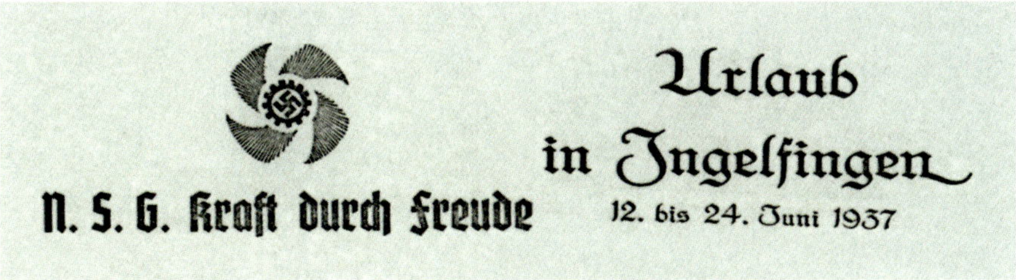

The National Socialist organisation 'Strength through Joy' advertises holidays in Ingelfingen (1937)

company that made harnesses for oxen. The district creamery, which was established in the late nineteenth century, and the Huber, Hofmann & Co. canning factory (also known as the 'food factory'), which was founded in the town in 1921, both contributed to the economic upswing in the town. Moreover, thanks to the National Socialist 'Strength through Joy' holiday and leisure programme, Ingelfingen had been developing into a popular holiday destination since 1933.

The resulting economic upswing increased the need for manpower. When war broke out in September 1939 and all men who were liable for military service were called up, Ingelfingen suffered a shortage of workers. Initially, women filled many of the vacant jobs. Later, as was the case all over Germany, increasing numbers of prisoners of war and forced labourers were drafted to do work. A camp for prisoners of war was set up in the Ingelfingen Festival Hall. Together with forced labourers from the Ukraine, Poland, Russia, and Belgium, these POWs worked in companies and on farms in the locality. Female forced labourers from Poland were mostly drafted to work in the canning factory. It was here at the 'food factory' that Marie Müller soon found work on the production line. With the money she earned, she was able to rent a flat for herself and the children in 1942. This was the first step towards a simple new life.

Only a few months later, the war reached the Müller family in Ingelfingen. Starting in 1943, an increasing number of women and children were evacuated from nearby cities to the Kocher Valley. Quite literally bombed out of their homes, these evacuees had survived the first of the devastating air raids flown by the western Allies. This explains why, unlike many of the locals, they adhered to the black-out regulations and hid in the cellars as soon as the air raid sirens sounded from the roof of the town hall. Marie Müller was also terrified of the air raids. Her son Fritz recalls a number of nights in the final months of the war when he, aged six, and his mother and brother took shelter in the cellar of their house and waited for the signal that the coast was clear.

From early 1945 onwards, the evacuees were joined in Ingelfingen by countless refugees from Silesia and the Fichtelgebirge (a mountain range in Bavaria). Their presence, the dreadful state they were in when they arrived in the Kocher Valley, and the stories they told about their flight made the people of Ingelfingen even more war-weary than they were already. By the winter of 1944/45, hardly anyone in the town believed that Germany would win the war. In the autumn of 1944, they had followed the call to join the *Volkssturm*, a home defence force organised by the National Socialist leadership that mobilised all able-bodied men who were not already at the front for the defence of the homeland. However, the poor leadership of the *Volkssturm*, the fact that its volunteers were poorly armed and equipped, and the prospect of having to senselessly sacrifice one's life so close to the end of the war increased the desire of the people of Ingelfingen for an end to the misery and hardship.

This explains why most of the people who put up resistance to the approaching American troops in the region in early April 1945 were not actually natives of the Kocher Valley. Eye witnesses report that despite local protests, a *Wehrmacht* commando blew up Ingelfingen's bridges over the river Kocher on 8 April 1945. But the detonations could not halt the advance of the American troops. On 9 April 1945, the mayor of Ingelfingen surrendered control of the town to the Americans.

But the war was not yet over for the people of Ingelfingen; fighting continued in the surrounding region for another two days. Just before the Americans marched into Ingelfingen, the surviving members of a German battalion of cadets had taken cover in the woods to the north west of Ingelfingen. The young cadets had been commanded to hold the Kocher Valley and stall the advance of the enemy. But not all soldiers obeyed orders at this late stage of the war. Fritz Müller's ten-year-old brother and friends saw some German wireless operators scuttling their car and all their equipment in the river. Fritz Müller himself witnessed the execution of deserters by a military police commando on the outskirts of Ingelfingen. The remaining soldiers in the battalion of cadets tried – whether out of conviction or out of fear for their lives – to stop the advance of the American troops. These German positions came under heavy American fire for two days. Hardly any German soldiers survived. There were also a number of civilian victims on homesteads outside Ingelfingen.

Initially, the people of Ingelfingen welcomed the arrival of the Americans as a 'salvation from misery and hardship'. This soon changed, and they began to view the US soldiers as nothing more than an occupying force. The soldiers did nothing to stop former prisoners of war and forced labourers from looting and stealing in the town. It even seemed as if they sometimes encouraged these violations, considering them to be a kind of compensation for the injustices suffered by these so-called 'displaced persons' in Germany. The people of Ingelfingen, however, were barely conscious of this injustice and expected the Americans to protect their property.

Something that upset the locals even more, however, was the fact that the US troops requisitioned private homes to accommodate their men. During the war, the number of people living in Ingelfingen had risen from about 1,200 to over 1,500 as a result of the evacuees and the refugees from the East. This meant that flats and houses in the town were already overcrowded before the Americans marched into Ingelfingen. When the American troops took over, up to 800 American soldiers had to be billeted in the town. For this reason, a number of locals were obliged to move out of their homes in April and May 1945 in order to make room for the families of American troops.

Marie Müller was no exception. The American soldiers gave her half an hour to pack the family's most important belongings and to load them onto the wooden cart that her brother Christian had used to transport the small family and its belongings to his home a few years previously. The elegant walnut furniture that Marie Müller had brought to Ingelfingen as a reminder of the years spent with her husband in Esslingen, had to be left behind. Fritz Müller and his brother ran to their old home every day to take a look inside. What the two boys saw, horrified them: not only were the carefully tended frogs that they had raised in a homemade terrarium dead, what was much worse was that the 'half wild Yankees' had put a keg of beer right in the middle of their mother's good coffee table. 'There were whiskey bottles

Marie Müller (left) with her sons and her brother's family in front of her flat on Bühlhofstrasse (c.1944). After the arrival of the Americans, Christian Hasel's wooden cart was used as a removal cart whenever Marie Müller had to move out of her flat to provide accommodation for members of the occupying forces

everywhere; their weapons stood in the cupboards. It was all very distressing for my mother,' recalls Fritz Müller.

By early August 1945, most of the American troops had been withdrawn from Ingelfingen and Marie Müller and her children were able to return to their flat. However, they did not stay for long. Fritz Müller recalls that in the years that followed, the family had to vacate the flat several times to make way for American soldiers. The repeated experience of suddenly being driven out of one's own home as well as the uncertainty and the feeling of homelessness that dominated the weeks before they were allowed to return to a flat in which soldiers had been living, left indelible marks on Marie Müller.

The situation was compounded by the economic hardship of the months immediately following the end of the war. The Americans confiscated and took away not only all of the stocks in the canning factory, but also the tin for the cans too. It was months before production got underway again. Along with the other workers in the factory, Marie Müller was permanently laid off. Once again, she was dependent on the support of her brother.

When production started up again at the canning factory, Marie Müller's wages were just enough to buy what was absolutely necessary. Toys or sweets for the children were an unattainable luxury; clothes had to be above all durable. In the years that followed the war, Fritz Müller possessed little more than a pair of lederhosen. 'My brother and I went to school in lederhosen – no shirts, no nothing,' he recalls, 'we wore those lederhosen when we went swimming in the river Kocher in the summer and when we went to bed at night.' Their shorts may have been made of leather, but leather shoes were completely unattainable. In the summer, Marie Müller was forced to send her sons barefoot to school. In the winter, they wore wooden clogs that a refugee from the Erzgebirge (a mountain range in the east of the country) had made in exchange for food.

In the years immediately following the war, the focal point of life was getting food. During the war, the ruthless exploitation of the territories

To school in lederhosen: Fritz Müller (back row, far right) in a class photo from the summer of 1948

occupied in the East and the targeted rationing of food made sure that the German population didn't go hungry. With the defeat of the Third Reich, this system collapsed. The Americans, the Britons, and the French did make an effort to supply the Germans with the bare necessities. However, two factors affected their efforts: firstly, almost the entire population of Europe was dependent on the food provided by the Allies and secondly, the damaged infrastructure hampered transportation. In the first two years after the war, food shortages were such that the rations printed on the food ration cards were not always available for distribution. Even in the American zone of occupation, which was comparatively well supplied with food, daily rations provided just over 1,300 calories – less than the target 1,500 calories. In the other occupied zones, the calorific value of daily rations was much lower.

Many Germans went hungry. The only ones who didn't were those who were self-sufficient and farmers who were able to sell their produce on the quiet. As the black market boomed, the Reichsmark rapidly devalued.

People earning small wages, like Marie Müller, were the first to be hit by this state of affairs. Like many Germans who drove out to the country to exchange whatever they felt they could do without – clothes, cigarettes, jewellery, or anything else of value – for food, she and her family were only able to ward off starvation by bartering.

As a local-born woman, she was in a better position than the evacuees and refugees in Ingelfingen because the farmers knew her and sometimes slipped her a piece of bacon or a packet of butter. Marie Müller was grateful for these gestures, but found it embarrassing to be reliant on the kindness of people that she had known since her childhood.

Kurt and Fritz Müller helped their mother in whatever ways they could. They caught fish in the river Kocher and ran along behind the combine harvester collecting any grains of corn that had been left on the ground. In the autumn, they went into the woods armed with brooms and sieves to collect beech-nuts that could be ground in the local oil mill. The forest, explains Fritz Müller, was 'as clean as a whistle. There wasn't a twig to be found anywhere. In fact, there was absolutely nothing to be found at all.' Those who were able, gathered and stored a stockpile of wood for heating and cooking because coal and other fuels were in short supply. The winter of 1946/47 was the harshest in 20 years. When the temperatures dropped as low as 20 degrees below zero towards the end of the year and the supply of coal throughout Germany collapsed because of transportation problems, the American authorities in Ingelfingen enlisted all able-bodied men between the ages of 18 and 60 to work in the woods to ensure a supply of fuel for the population.

In 1946, the Müller family spent the holiday season in an unheated flat. 'We had almost no fuel that Christmas,' recalls Fritz Müller. 'My mother had a little stove with three or four hotplates and an oven in the kitchen. She fed it with some wood and the three of us put our feet in the oven to keep warm. A Christmas tree – or anything like that – was completely out of the question.' The daily struggle to survive, the experience of having to rely on the support of others despite all her efforts, and maybe even the worry that she was not being a good mother to her two boys put a real strain on Marie Müller. Despite the fact that she could always rely on her brothers and her old friends in Ingelfingen and that she derived strength from her efforts to bring up her boys well, ultimately she alone was responsible for the children. It is possible that this extreme burden was the reason why Marie Müller clung for so long to the hope that her husband had not in fact been killed in action, but was just missing or a prisoner of war and that he would one day return to her and her two sons. In 1941, her husband's identification card, identity tag, wedding ring, and other personal items had been returned to her along with the official notification that Fritz Müller Sr. had been killed in action. Nevertheless, her son confirms that for years, Marie Müller sat and listened to the Red Cross's missing persons reports on the radio every day. It was only several years later, when a soldier who had served in the Balkans with

Hohenlohe c.1950: just like these children, in the years that followed the war, Fritz Müller and his brother Kurt scoured the empty fields for grain that had been left behind by the farmers

her husband contacted her and told her that he had seen with his own eyes how Fritz Müller had been shot dead by a Russian sniper and knew where he lay buried, that Marie Müller stopped hoping that her husband would one day return home.

It is likely that her faith gave her the strength she needed to make it through these difficult years. In the 1920s, Marie Müller's mother joined the New Apostolic Church. A short time later, Marie Müller's brother Christian followed suit. After the war, Marie Müller and her sons also joined. Looking back, Fritz Müller assumes that his mother was attracted more than anything else by the close-knit church community where she, as a widow raising two children, found support and friendship. Marie Müller remained a member of the New Apostolic Church until her death in 1965; her son Fritz is to this day an active member of the New Apostolic community in Ingelfingen.

The family's financial situation stabilised in the late 1940s. The currency reform in the summer of 1948 put an end to the years of shortage and black market dealings. The 120-Mark surviving dependents' pension that Marie Müller received every month for herself and her sons was the backbone of the family's modest income. She supplemented this income with the wages that she earned as a worker in the production unit of the companies Reisser and Bürkert, which were established in Ingelfingen after the war. In 1950, Fritz Müller's elder brother, Kurt, began commercial training in the nearby town of Künzelsau and was able to contribute to the support of his mother and younger brother. Two years later, the family laid the foundation stone for their own house in Ingelfingen. The years of hardship were over. Germany's 'economic miracle' was about to begin.

CHAPTER 3

MAKING SOMETHING OUT OF NOTHING: FRITZ MÜLLER'S CHILDHOOD AND YOUTH IN THE KOCHER VALLEY (1942–53)

'After my father's death, the thing that mattered most to my mother was the happiness of her children,' says Fritz Müller. 'Everything she did, she did for us.' Marie Müller's efforts were obviously not in vain: when Fritz Müller talks about the years of deprivation and uncertainty both during and after the Second World War, he makes them sound like a period of high adventure. Whether it be nights spent huddled together in the cellar during air raids or days spent collecting empty cartridge cases after low-flying aircraft attacks, playing dangerous games in the forest or on the banks of the river, or taking part in soap-box derbies, Fritz Müller paints a picture of a happy and exciting childhood. The children hardly noticed that they lacked anything after the war. 'Apart from the sons of the canning factory owner, we were all the same,' says Müller, looking back, 'none of us had anything.'

This was a fate they shared with most people in Germany. In the final years of the war, National Socialist propaganda encouraged citizens to collect and re-use materials such as paper, clothes, and metal. In the months following the end of the war, when there was hardly anything in the shops to buy, the Germans had no choice but to resort even more frequently to recycling and improvisation. Each of them tried to make something useful out of the little they had. In 1946, for example, Christian Bürkert, an Ingelfingen-born engineer, used the remains of a fighter-bomber that had been shot down to make simple incubators, foot-warming devices, and stoves, thereby laying the foundation for what would later become his thriving business. Others made new pieces of furniture out of old bits of wood, used shell cases as coffee machines and steel helmets as kitchen sieves, or transformed old petrol cans into watering cans for their vegetable patches. For their part, the women made new dresses out of old. The youth magazine published by the German Confederation of Trade Unions in 1948/49 even provided tips on how to knit a bathing suit out of leftover wool and to make a bag in which to carry it using sawn-off broomsticks and sackcloth.

Children also improvised and made things themselves. Says Fritz Müller: 'We made our own toys. A little piece of wood and four wooden discs as wheels was all we needed to make a car.' He also made little wooden water-wheels and played with them at the river and used empty cans of

The war forced the Germans to improvise: a young man in post-war Berlin uses his mother's sieve to water the allotment

During the school holidays, Fritz Müller (seen here with his mother) often visited his paternal grandfather, Friedrich Müller, in Esslingen

long-life meat and bread discarded by the Americans to make simple tin toys. 'First I cut open the cans. Then I used a nail to bore holes in the tin and joined the pieces up with wire or small screws to make cranes, aircraft, and, of course – it was, after all, just after the war – tanks, and canons.'

In the summer months, Kurt and Fritz Müller often visited their grandfather in Esslingen, where they played in the stock yard among the sand, plaster, and tools, and watched the workers slaking lime. Sometimes, when they knew no-one was watching, they would sneak into their grandfather's garage and sit in his Mercedes-Benz. Müller recalls: 'I can well remember the day my brother and I managed to start the car. The engine sprang into life and we burst out through the closed garage door. There was, of course, all hell to pay. We were both given a good thrashing.'

When they were not in Esslingen, Kurt and Fritz Müller played with other Ingelfingen boys in the forest or at the river. The adults worked during the day and hardly had time to look after the children. As a result, the Müller boys and their friends could virtually do whatever they liked. Although Kurt, who was just under eleven years of age when the war ended, had been instructed to look after six-year-old Fritz, he didn't always want his little brother tagging along behind him. 'I remember once my brother tied me to a tree in the forest and went off to play with his friends for a couple of hours,' says Fritz Müller.

Marie Müller must have often worried about her son's escapades. The fact that they stole eggs from nests – something she strictly forbade them to do – as harmless in comparison to some of the other antics they got

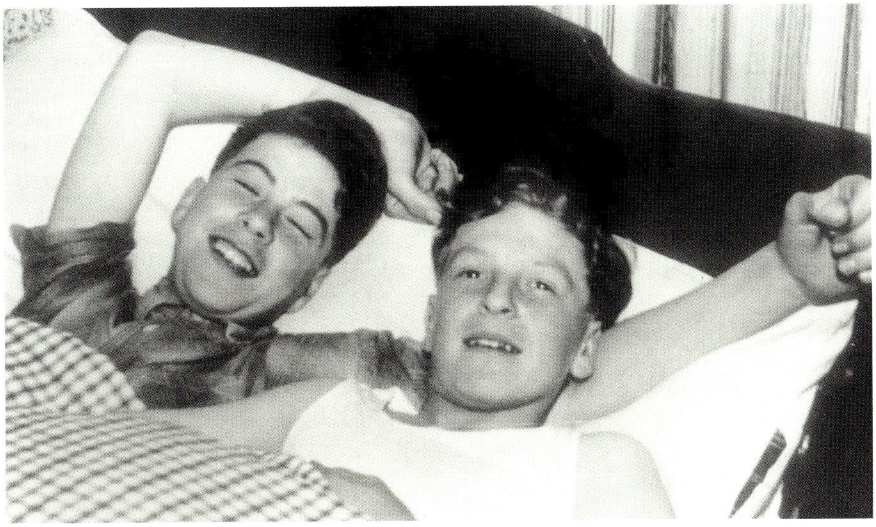

Fritz Müller (left) and his cousin Heinz were inseparable as children

up to. It was much more exciting for the boys to play with the weapons and ammunition that had been abandoned in the woods around Ingelfingen when hostilities ceased. 'Wherever soldiers passed through an area, they abandoned cars, machine guns, pistols, and radio equipment,' recalls Müller. 'The woods were full of such things. Naturally, we found this thrilling.' The boys shot targets with the abandoned machine guns and used hand grenades to catch fish in the river. 'The best and easiest way to catch fish was to use hand grenades,' says Müller. 'Pull off the ring, chuck it into the river, and BANG! Lots of dead fish float to the surface.' Most of these games passed off without serious consequences: a few scratches here, a broken arm there. Sometimes, however, the boys sustained much more serious injuries. One of Fritz Müller's playmates, for example, lost an arm when a hand grenade exploded while he was playing with it. But instead of frightening the boys, the risk merely heightened the attraction of such dangerous activities.

The boys of Ingelfingen also enjoyed regular scraps with the boys from the neighbouring Jagst Valley. However, they were never short of a reason to fight other kids from Ingelfingen either. Depending on who was involved, it was a case of Upper Town against Lower Town, Old Town against New Town, or locals against refugees.

Looking back, Fritz Müller describes his childhood as a 'completely wild time for my brother, myself, and other youngsters of my age.' During these wild years, the kids also stole cans of food or other useful items from the trucks of the American army. 'Of course the GIs saw what we were up to,' says Fritz Müller. 'But they also saw that we had nothing. That's probably why they didn't punish us. If they caught us, the worst that happened was that they would get us to start fighting each other. The soldiers enjoyed watching us scrapping.'

The American GIs might have enjoyed watching young Germans boxing and wrestling each other, but the American military government was very worried that these half-wild children would grow up to be rowdies with a propensity to violence. They were even more worried about those young people who had received a semi-military training in the Hitler Youth or its subdivision, the German Youth. This is why in the autumn of 1944, the US Army drafted plans for activities for children and young people that would emphasise the ideals of democracy and freedom. The plan also foresaw that German trade unions, the Churches, and other civilian groups would get involved in youth work again as quickly as possible.

These plans, with which the Americans sought to establish an independent system of youth work in Germany without American involvement, were, however, undermined by the fact that the American soldiers in Germany did not always obey orders from military command to keep all contact with the German population to an absolute minimum. The GIs might have been able to treat German adults with indifference, but not so the small children 'who always crowded around the soldiers whenever convoys stopped in villages or towns. Many of the soldiers loved seeing the joy that spread over the kids' faces when they were given sweets, chewing gum, or something to eat from the soldiers' military rations. These little presents were soon followed by initially tentative attempts to talk to or play with the kids,' notes an official US Army report.

This is why the GIs were permitted to have contact with German children from early June 1945 onwards. Soon afterwards, the US soldiers began spontaneously organising leisure activities for German children and young people. This was not what the military government had originally intended when it planned activities for German youths. In July 1946, a compromise was achieved with the establishment of the 'German Youth Activities Section' (GYA). From that point on, members of the US Army were allowed to help German civilian groups set up activities for children and young people in order to introduce young Germans to democratic values through games and sport. The GYA operated for just under ten years. During this time, it funded sporting events, sports courses, DIY, and handicrafts, as well as orphan-

First steps in international reconciliation: American soldiers and German children in the summer of 1945

ages and homes for young people. It also organised excursions, film showings, dances, Christmas parties, needlework competitions for girls, and a 'Master Singer Competition' for young singers. The most popular GYA initiatives were, however, the soap-box derbies.

While self-made, motorless soap-box carts were not unknown in Germany, they were certainly by no means as popular as in the USA, where the 'All-American Soap Box Derby' became one of the most important national sporting events in the 1930s. In the spring of 1947, American soldiers in Ulm organised the first soap-box derby. Similar events all around the American-occupied zone soon began attracting tens of thousands of onlookers. Initially, the winners were rewarded with CARE packages. Later on, the prizes included radios, bicycles, and other valuable non-cash prizes. Germany was gripped by soap-box fever! In 1949, just under 15,000 boys between the ages of ten and fifteen took part in over 500

Learning democracy the fun way: in the early 1950s, boys between the ages of 10 and 15 all over the Federal Republic of Germany worked hard on their soap-box carts

soap-box derbies in the American sector. Just like in the USA, qualifying races were held at local and regional level. The winners qualified for the grand finale race which took place on the Theresienwiese in Munich on 24 July 1949. Some 12,000 people came to watch. The champion won a two-week trip to the USA. In 1950, 20,000 young boys from all over Germany took part in the qualifying races. Weekly cinema news bulletins reported on the races and a number of children's books telling how creative young boys became the heroes of major and minor derbies alike were published (*Willy wins the race* (1951), *The Super Soap-box* (1952) or *Make Way for Flix* (1953)).

Kurt and Fritz Müller were also bitten by the bug and dreamed of racing fame. Together with one of Kurt's older friends, they built their own soap-box cart. Because the Müller boys could not afford expensive materials like plywood for the body and metal for the undercarriage of their cart, they had to improvise. Both loved to make model airplanes in their spare time and were able to apply the design principles of aircraft construction to their soap-box cart. They used strips of wood to make a lightweight frame, which they then covered with material, namely Marie Müller's black-out curtains from the war. They also cut up an old sheet to make white stripes so that their cart would look more like a racing car. The aerodynamic body and the seat, which was covered with red material, were placed on top of an undercarriage made of slats and featuring pram wheels. A wooden beam into which rubber had been nailed served as the rear wheel brake, which all soap-box carts had to have. The finishing touch was a silver-painted cardboard Mercedes star, which the boys attached to the cart's 'radiator grille'. The Müller's soap-box cart was now ready for its first test-drive through Ingelfingen. The boys had told everyone they knew about this important event. Given the widespread enthusiasm for soap-box derbies, it came as no surprise that half the town turned out one spring Sunday morning in 1948 to watch Fritz Müller hurtling downhill in his homemade cart.

Encouraged by the success of this test-drive, the three young 'mechanics' entered the next official soap-box derby in Langenburg. This race showed

Fritz Müller (second from right), his hand resting leisurely on the steering wheel of his soap-box cart, before the race ...

... and (third from right) during the award ceremony afterwards with his prize, a Bilora box camera, in his lap

that Fritz Müller was the ideal weight for the soap-box. 'I started very well from the ramp and then accelerated even more. This helped me finish at the front,' he recalls. Although his competitors' heavier soap-box carts picked up speed towards the end of the race, they were relatively slow to get going at the start and were playing catch-up from the word go. His own brother, who was a few kilos heavier than Fritz, didn't chalk up a win.

Having won this first race, Fritz Müller finished near the front of the field at the subsequent district qualifiers. During the regional qualifiers in Stuttgart, however, disaster struck. Here, he did not race in his own cart, but used a cart that the Ingelfingen entrepreneur Christian Bürkert had had made for him. Bürkert had done so because Fritz was not only successfully representing Ingelfingen at the races, he was also the son of an employee, Marie Müller. Even though this new vehicle was equipped with coveted Opel ball bearing wheels, which Fritz and his brother could never have afforded for their homemade cart, it was too heavy. The problems began when the drivers and their carts were weighed on the eve of the race: the undercarriage of the vehicle that the well-meaning sponsor had made for the lightweight driver was so solid that the vehicle exceeded the maximum permissible weight for the cart. Fritz and his brother spent the night before the race removing the heavy wooden base and replacing it with lighter slats.

But it made no difference. Even after replacing the undercarriage, Fritz and his wooden monster stood no chance of winning the race: he was up against competitors who had built their soap-box carts with the support of the testing department at Daimler-Benz. In their streamlined, metal and aluminium carts, which had been specially designed to suit their weight, they sailed effortlessly past Fritz Müller's heavy cart, which accelerated much too slowly and only reached its maximum speed towards the very end of the track. As Fritz Müller sped into the bales of straw at the finishing line, he knew his high-flying dreams of taking part

26

in the grand finale in Munich lay in tatters. To this day still disappointed by what happened, he says: 'If I had taken part with my own cart and had replaced its pram wheels with the Opel ball bearing wheels, I would have stood a real chance. But with that solid thing there was absolutely nothing I could do. It was just hopeless.'

Even though Fritz Müller never entered another soap-box derby after this defeat, his enthusiasm for making and designing remained as strong as ever. He was very skilled in dismantling old motorbike engines, repairing them, and putting them back together again. He later concentrated on creating airplane models that could fly.

Building model airplanes had been popular in Germany since the late 1920s. The National Socialist regime fostered the enthusiasm for flying by giving young boys the chance to fly gliders in the aviation division of the Hitler Youth. In the post-war years too, building model airplanes was popular in Germany and was widely encouraged because it was considered an enjoyable way for children to learn about thermals and aerodynamics – the ideal starting point for mechanical/technical professions. In 1954, the design engineer Werner Weber wrote the following in his Construction book for father and son: 'The creative drive that is inherent in every human being finds an outlet in constructing models. It is often during the hours spent in this manner that a child discovers the talents and inclinations that can shape his choice of profession and later life.'

This was certainly the case with Fritz Müller. Having built his first simple glider models, he began to design motor-driven models. Soon the twelve/thirteen-year-old began making his first self-made single-channel radio-control system. 'I could only control the rudder with this system,' he recalls, 'but at least it worked.' But that was

The soap-box derbies, which were sponsored by Opel, attracted large numbers of onlookers well into the 1950s

only the beginning. As time went on, his models became more complex and more sophisticated. Fritz Müller adored projects such as these and decided to make his hobby his career. He decided to become a precision engineer.

Fritz Müller with his self-made model airplane on the banks of the river Kocher in the early 1950s

CHAPTER 4

APPRENTICESHIP AND EARLY PROFESSIONAL EXPERIENCE (1953–63)

At the age of 14, in the summer of 1953, Fritz Müller obtained his elementary school leaving certificate. Because his mother did not have enough money to pay the school fees needed for secondary school, he had already begun looking for a post as an apprentice in the spring of that year. Although the boys' grandfather, Friedrich Müller, had always promised that Fritz and his elder brother Kurt would be heirs to his plastering business in Esslingen, Müller's sons-in-law – who had joined the company after their release from prisoner-of-war captivity in 1945/46 –now claimed precedence as heirs to the company. In 1949, when Kurt Müller was due to start his apprenticeship in his grandfather's business, an acrimonious dispute broke out in the family. Marie Müller decided that her sons should not have to rely on the graciousness of their relations in Esslingen, but should instead stand on their own two feet. A short time later, Kurt Müller began commercial training in the nearby city of Künzelsau.

Fritz Müller also wanted to complete his apprenticeship in the Kocher Valley. On the surface of things, the prospects were good. The number of employed people in Ingelfingen rose from just under 500 to approximately 840 between 1950 and 1955. Manpower was needed, especially in the construction industry. Just like Marie Müller and her sons, many people in Ingelfingen around this time bought an inexpensive plot of land from the town and took advantage of the low-interest loans on offer from

The 'economic miracle' triggered a building boom in Ingelfingen: Marie Müller and her sons were just one of many families to realise their dream of their own home in the early 1950s

the state to realise their dream of building a home of their own. The district housing association in Künzelsau built numerous blocks of flats and many companies extended their premises. Large companies like R. Stahl, the Stuttgart-based manufacturer of materials handling systems, Ziehl-Abegg, the engine and drive manufacturer, Sigloch bookbindery (which had production units in Künzelsau since the mid 1940s), the screw manufacturer Reisser, and many more were expanding and constantly increasing their workforces.

The other companies in Ingelfingen and the surrounding region also benefited from the economic miracle. After the lean post-war years, the products sold by the district creamery, the preserves, marmalades, and jams made by Ingelfingen's 'food factory', and local wines such as the 'Schloßgeist' and the 'Criesbacher Kocherperle' were going down a treat with consumers. Another locally manufactured product was also taking the German market by storm: riveted, American-style jeans. In 1948, the managing director of a small Künzelsau-based clothing company, Kleiderfabrik Hermann, persuaded an American GI to exchange six bottles of home-made schnapps for six pairs of jeans, which he used to make patterns for his own production unit. Both the blue jeans and the girls' camping trousers, which were launched on the market in 1953, sold like hot cakes. The Mustang brand was established in 1958 and fashion-conscious teenagers across the country were soon buying several pairs of jeans at a time.

Well-fed and well-dressed as they now were in the 1950s, the Germans wanted to see something of the world. Those who could not afford a dream holiday in Italy or found the journey south too long or arduous, were certainly not short of relaxing holiday destinations in Germany. 'Bad Ingelfingen' (Ingelfingen Health Spa), which began offering cures again in the 1950s, was one such a holiday destination. With its medicinal water, which was recommended to those with stomach or liver complaints, 'Bad Ingelfingen' promised relief from a variety of diet-related complaints suffered by many Germans during the economic miracle.

Ingelfingen's medicinal water promised relief from the health complaints associated with the abundance of food available during the 'economic miracle'

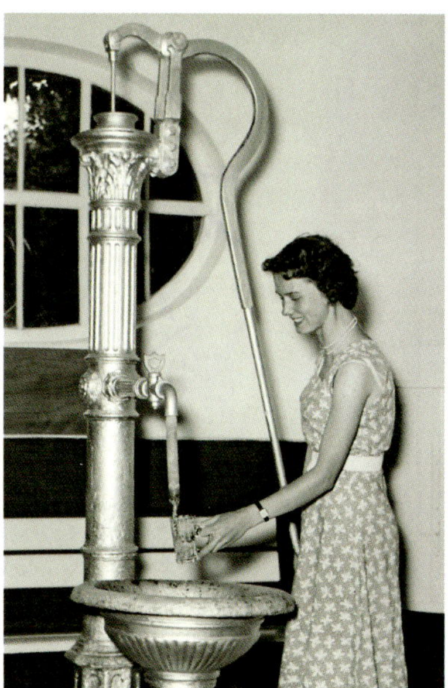

However, although Ingelfingen's economy was strong in the 1950s, there was a shortage of apprenticeship positions. Only a few skilled workers were needed on the construction sites because most future home-owners rolled up their sleeves and did much of the work on their homes themselves in the evenings and at weekends. Even the spa sector and industry preferred to hire unskilled workers because of the lower wages involved. It was only the small businesses in the trade sector that took on apprentices, albeit generally only as cheap temporary labour for a variety of odd jobs.

'In these little one- or two-man businesses, the apprentice was really nothing more than a "gofer",' recalls Fritz Müller. 'The gofer had to get the bread for lunch, sweep the workshop, run errands for the boss's wife, and help in the garden.' The variety of apprenticeship positions on offer was limited. 'There was a carpenter, where you could learn to become a carpenter; there was a plumber, an electrician, and a couple of other small businesses that mostly delivered or repaired equipment and machinery for wine-growers ... that was about it really,' says Müller.

This meant that Fritz Müller's prospects of getting an interesting apprenticeship position that was in line with his ambitious plans were poor. The fact that he was able to actually train to become a precision engineer in his home town was all due to Christian Bürkert (1915–71), the son of an Ingelfingen farmer, who hailed from Bühlhof, a hamlet near Ingelfingen. Having finished school, Bürkert completed a technical apprenticeship at a large electronics company in Berlin before studying engineering.

Not much on offer: in the early 1950s, there were only a handful of trade businesses in Ingelfingen and a correspondingly small selection of trades that could be learned (photos from the 1949 Ingelfingen trade fair)

The Bürkert company stand at the Ingelfingen trade fair in 1949

At the end of the war he returned to Ingelfingen and established a company in the basement of his parents' farm. His very first product was a combination of his knowledge of agriculture and his ability as an engineer. Using the aluminium casing of a fighter-bomber that had been shot down and the sheet iron of an empty vat of chemical fertiliser, Bürkert constructed a bimetal temperature regulator for incubators that allowed a large number of chickens to be hatched quickly and reliably at the same time. In the lean post-war years, Bürkert did a brisk trade in his simple yet robust incubators (or hatcheries). He also installed his temperature regulators in foot-warming devices during the bitterly cold winter of 1946/47. As electric domestic appliances became more widespread in the 1950s, Bürkert supplied a number of oven, fridge, and washing machine manufacturers with his temperature regulators. Soon he was producing washing machine valves and oil regulators too. As the years passed, he extended his range of products for the control and regulation sector. His business was growing and growing.

In order to ensure his company's continued success, Christian Bürkert needed well trained, highly skilled workers. And so in 1951, when he had to move to a larger factory building for the third time in five years, he converted his old production unit into a workshop for trainees. Every year, about ten young people from Ingelfingen and the surrounding region completed a three-and-a-half-year apprenticeship to become a metal-worker, toolmaker, or precision engineer at Bürkert's workshop. This short yet intense apprenticeship, which was overseen by a master whom Bürkert had hired from a large Berlin-based company and brought to Ingelfingen especially for this purpose, was unique in the district of Künzelsau.

This meant that Bürkert was the best company for apprentices in the Kocher Valley and the ambitious and budding engineer, Fritz Müller, was determined to become an apprentice there. Wanting a position was one thing; getting one was quite another. Young men from all around the region applied for the handful of apprenticeship positions on offer at Bürkert. What's more, Fritz Müller was born between 1938 and 1942, a period of high birth rates in Germany – the outcome of National Socialist birth policy. Appeals made by politicians for businesses to make a sufficient number of apprenticeship positions available for these young people had largely fallen on deaf ears: in the early 1950s, there was a shortage of apprenticeship positions in the country. This was certainly true of Ingelfingen and the surrounding region. When Fritz Müller applied to Bürkert in 1953, there were six applicants for every available position. Each of the applicants had to sit a tough entrance exam where the examiners paid more attention to the applicant's technical skills than their academic abilities. Fritz Müller's experience building soap-box carts and model airplanes stood him in good stead during this exam. What's more, the young man's mechanical skills had come to Christian Bürkert's attention five years previously during the soap-box derbies. Nevertheless, Müller fretted for days about whether he would get the position he so dearly wanted at Bürkert. When on 10 April 1953 his mother finally received written confirmation that he had been accepted by Bürkert, mother and son were over the moon.

Having finished school, Fritz Müller got a summer job in a tree nursery near Waldzimmern. In September 1953 he began his apprenticeship as a

Betrieb	Buchh.	Einkauf	K.B.	Pat.	Pers.	Verkauf	Verwalt.		
		4. APR. 1953							

Erledigt am:

An die Mutter
des Schülers Fritz Müller

(14a) Ingelfingen/Württ.

b/St 8. April 1953

Lehrvertrag.

Wir sind bereit, Ihren Sohn Fritz zu uns in die Lehre zu nehmen.

Zur Vorbereitung des Lehrvertrages wären wir Ihnen dankbar, wenn Sie auf der beiliegenden Copie dieses Schreibens uns die nachstehend genannten Fragen beantworteten und uns die Copie zurückgäben.

Name des Lehrlings: Fritz Dieter M ü l l e r

Geburtsdatum: 2.1.39 Geburtsort: Esslingen Neckar Wohnort: Ingelfingen

Gesetzlicher Vertreter (Vater, Mutter, Vormund): Marie M ü l l e r

deren Wohnort: Ingelfingen / Wttbg.

Abgangsklasse und Schule: 8. Klasse

Beruf des Erziehungsberechtigten: (Hausfrau)

Wir empfehlen Ihnen, sich bei Ihrem Bürgermeisteramt schon immer eine Steuerkarte und eine Invalidenkarte für Ihren Sohn zu beschaffen.

Den genauen Termin des Lehrbeginns können wir wegen notwendiger baulicher Veränderungen in unserem Hause Ihnen noch nicht nennen. Wir werden voraussichtlich im Monat August d.Js. den neuen Jahrgang bei uns aufnehmen.

Hochachtungsvoll
Christian Bürkert
Bau elektrischer Geräte

1 Copie anbei.

precision engineer. The apprenticeship was tough. Particular attention was paid to cleanliness and tidiness. Anyone who turned up for work with a greased quiff was sent straight to the barber's. Every evening, the master checked to make sure that all the tools had been properly tidied away. Any carelessness was punished with a week of chores such as sweeping the workshop or cleaning the toilets. But despite the strictness of the apprenticeship, the quality of the training at Bürkert was excellent. 'As far as the training was concerned, at Bürkert

Fritz Müller was overjoyed to receive confirmation that he was to become an apprentice at Bürkert

Fritz Müller (front left) as an apprentice in the apprentices' workshop at Bürkert

I got the best that was available,' says Müller. He was eager to learn and could hardly get enough. 'The master had to throw me out of the workshop at 7.30 p.m. I would still be in there at that hour working on gears, or a watch, or something like that. I was completely in my element,' he says.

Fritz Müller also felt very much at home at the vocational school. He was particularly fascinated by his teacher's little workshop, which joined onto the classroom. 'It had everything from universal joint shafts, electric motors, and dynamos to generators and broken nails. I loved going in there. As far I was concerned, this workshop was a little piece of heaven.' Both Müller's vocational school teacher and master encouraged the ambitious young man and made it possible for him to reduce his apprenticeship by six months to only three years. His teacher obviously thought very highly of him. Says Müller: 'I probably shouldn't be telling you this, but I never actually sat a theoretical exam at the vocational school. When it came to the final written exam, my teacher looked at me and said: "You know it all anyway." I was allowed to go and sit in his workshop. Through the workshop window, I watched the others racking their brains for the answers to the exam questions.'

Despite all this, Fritz Müller was certainly not a swot or the stay-at-home type. He still spent most of his free time with his friends from the town. When rock'n'roll reached the Kocher Valley in the mid 1950s, Fritz Müller never missed a hop – despite the fact that he was not himself a dancing man. Enjoy them as he did, his conscience always nagged at him because the New Apostolic Church was at the time strictly opposed to hops and other worldly pastimes. 'When I went to the cinema for the first time,' recalls Müller, 'I fully expected to be struck by the wrath of God.' While nothing untoward happened at the cinema, other episodes in Müller's life around this time were not so painless. The adults in Ingelfingen's old, established families kept their distance from the German refugees who had come to the town after being expelled from the former German territories in the East. They also tried to prevent relationships developing between their children and those of the refugees. 'Anyone

Teenagers in Ingelfingen in the mid 1950s

who associated with refugees was a persona non grata,' recalls Müller. When he himself fell in love with a girl from a family of refugees, his family was appalled. His elder brother, Kurt, did everything to keep the two apart. Eventually the girl broke it off with Fritz Müller.

Fritz Müller escaped the suffocating small-town atmosphere by throwing himself into the world of gliding. For him, the gliders were the next logical step up from his model airplanes and as such, it was only natural that he was one of the founding members of Ingelfingen's gliding club. Christian Bürkert, himself an aviation enthusiast, had paved the way for the foundation of the club in the mid 1950s when he made a plot of land on Lipfersberg near Ingelfingen available as a glider airfield. Together with other club members, Fritz Müller put things in order at the Lipfersberg airfield. Before he had turned 16, he got his glider's

licence without ever telling his mother he intended to do so. When she found out, she tried forbidding him to fly. It was no use, however. From the moment he had his licence, Fritz Müller spent every spare minute with his friends at the airfield or in one of the club's self-made gliders.

Earning his wings: 15-year-old Fritz Müller before …

… and after his first solo flight

Having successfully completed his apprenticeship, Fritz Müller was hired by Bürkert. He spent the first eighteen months working in the laboratory before being promoted to the design office. Bürkert had started producing valves in the mid 1950s and Fritz Müller was tasked with focussing on new valve types. Christian Bürkert monitored his progress closely and frequently praised him to his face. Müller was filled with pride whenever he heard the words 'Fritz, you are worth your weight in gold!' from his boss. He knew that although Christian Bürkert could be very difficult, he was also a great inventor and a good designer with many ideas and the necessary drive to put his ideas into practice.

In his spare time, Fritz Müller and one of his friends repaired defective incubators dating from Bürkert's early years. 'The units did not last forever,' recalls Müller, 'and when they broke down, the farmers would knock on Bürkert's door and ask for them to be repaired. By this time, Bürkert was a small industrial company and there was no-one there who had either the time or the inclination to repair the filthy hatcheries.' Fritz Müller and his friend decided to take on the job: 'After all, the farmers had come all that way with their apparatus and could not understand that no-one wanted to do anything about them.' The boys did not invoice the farmers on the basis of the time they spent repairing the machines, but instead on the basis of the estimated wealth of the farmer. 'If we thought that a farmer had money, we asked him for three times the actual repair cost. But if poor George from down the road arrived with his incubator, we'd repair it for free because we knew that he had nothing and led a very simple life. So while we weren't earning big money, our repair service wasn't doing too badly at all, and Christian Bürkert never invoiced us for anything,' says Müller.

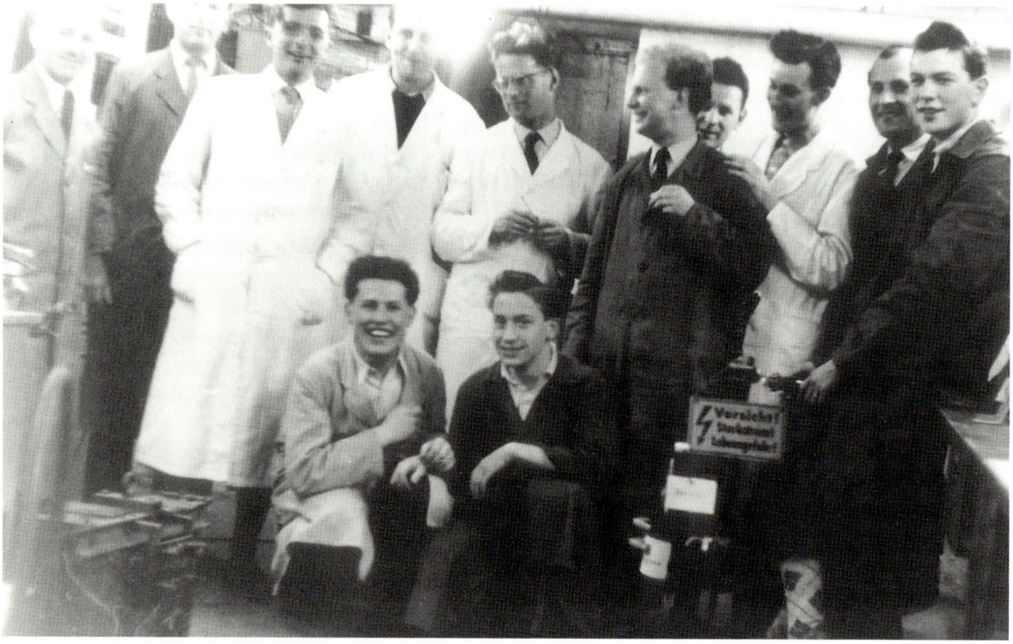

Having completed his apprenticeship, Fritz Müller (far right) spent eighteen months working in Bürkert's design department

Bürkert's well-meaning generosity was, however, quite capable of unexpectedly turning into resentment. Fritz Müller was at the receiving end of this resentment when he handed in his notice at Bürkert in 1959 to study precision engineering in Villingen-Schwenningen. When Christian Bürkert found out that Fritz Müller was still coming to Lipfersberg to glide at the weekends, he drove out to the airfield in person and gave Müller a dressing down in front of other club members. 'Bürkert was so furious,' recalls Müller, 'that he was barely able to put two words together at first. Once he recovered his composure, he told me in no uncertain terms that if I was no longer going to work at his company, then I was no longer welcome at his airfield.' Müller, who had given so much to the club and the airfield, was bitterly disappointed by this unjust treatment at the hands of his former boss. At the very least after this confrontation with Bürkert, the young aviation enthusiast promised himself that he would one day own his own aircraft.

The decision to study had not been an easy one for Fritz Müller. He had of late been earning good money at Bürkert and knew that if he did go to college, he was facing a number of lean years. 'I could have just continued working as a simple precision engineer,' says Fritz Müller. 'Many people back then did that in order to earn money straight away. But I wanted to learn more. I looked at the others and thought to myself: "Fine, you're doing pretty well right now and picking up speed, while I am taking it easy around the bends. But sometime soon, the road ahead of me will straighten out and I'll pick up speed and overtake you." And that is exactly what happened.'

Initially, however, it looked as if Fritz Müller wouldn't even make it past the starting line. He had set his heart on attending the precision engineering school in Villingen-Schwenningen in the Black Forest, not only because of the school's excellent reputation but also because, unlike many other technical colleges in Germany at the time, it was state approved. The entrance exam was correspondingly tough. Most of the other candidates had come from major watchmaking companies in the region, such as Kienzle or Junghans, to get a university qualification. They were much older than Müller and had more work experience. Fritz Müller felt hopelessly inferior to them. At the end of a full day of exams in Schwenningen, he just wanted to go home. 'My mother was waiting outside the school in our Isetta. When I came out of the exam, I looked at her and said: "Let's get in and go home. I've failed."'

FEINTECHNIKSCHULE SCHWENNINGEN a. N.
Staatliche Fachschule und Berufsfachschule für Feinmechanik, Elektromechanik und Uhrmacherei · Telefon 631

SCHÜLERAUSWEIS Nr. 2504
Name: Fritz Müller
geb. 2.1.1939
in: Stuttgart
ist Schüler des
1½-jährigen Technikerlehrganges
Eintritt in die Schule: 8.4.1959
Siehe Rückseite!
(Unterschrift)

Fritz Müller's student ID card for the precision engineering school in Schwenningen shows his carefully greased quiff

But he had underestimated himself. Only a few weeks later he received the letter of admission to the school of precision engineering and moved to the Black Forest. Money was tight. Even though his mother sent him DM 120 every month and he got money from his grandfather in Esslingen whenever he went to see him, he did not have enough money for a student flat of his own. Like many fellow students, Fritz Müller had to share a furnished room in the home of a Schwenningen family with another student.

'Our bedroom had previously belonged to the daughter of the family,' recalls Müller. 'We didn't have our own bathroom or toilet. The room was furnished with two beds, a wardrobe, a table, and two chairs. That was it.' The fact that Müller did not get on with his room-mate (they quickly ended up not talking to each other) did not improve matters at all. Thankfully, the comprehensive nature of the degree course he was taking left him very little time to worry about such matters. In addition to daily lectures in mathematics, physics, electronics, and precision engineering, there were practical classes, especially in the field of watchmaking. The students sat tough exams at the end of each semester. In short, the course was difficult, but Fritz Müller was ambitious and hard-working. In the summer of 1961 he passed his finals and became a state-certified precision engineer.

Back then, the job prospects for graduates of the precision engineering school were excellent. Fritz Müller only had to write two applications to get a job. Both the valve manufacturer Herion in Stuttgart and an electronics company near Tuttlingen invited him to job interviews and subsequently offered him jobs. Fritz Müller opted for Herion despite the fact that he had found the interview there very intimidating. He had been interviewed by a panel comprising the owner of the company, the head of the development department, and a senior commercial employee. Faced with such interviewees at the age of 22, he felt very small indeed. However, Herion was at the time the leading valve manufacturer in Europe and Fritz Müller, who was already determined to set himself up in business building valves someday, knew that he could gain valuable professional experience and learn all about state-of-the-art production technology at this Stuttgart-based company. He signed the contract with Herion and moved to Stuttgart.

The degree course at the Schwenningen school of precision engineering was challenging: this photo (c.1960) shows a seminar attended by Fritz Müller

Erich Herion founded the company Herion-Werke KG in 1938. There are striking parallels between the curriculum vitae of this entrepreneur and that of Fritz Müller. Born the son of an entrepreneur in Alsace in 1908, Erich Herion lost his father at the age of 11. Soon afterwards, the German family was driven out of Alsace when the region was returned to France after the First World War. Herion's widowed mother was completely destitute. Together with her eight children, she moved to Freiburg and spent the years that followed doing everything in her power to ensure they received a sound education. Erich Herion trained to become a precision engineer and soon worked his way up through the ranks to become manager of a company that manufactured control systems. However, because he knew that he could neither progress any further nor realise his ideas, he set himself up in business in Stuttgart in 1938 making solenoid valves for refrigerating plants. He manufactured these first valves in the basement of his home. They sold so well that Herion soon began renting a workshop. In 1942, he built his first factory building, which was destroyed two years later during a bombing raid. After the war, Erich Herion rehired his old staff and set up a new workshop for the production of valves in an outhouse that had previously been used to shelter sheep in Grossbettlingen near Stuttgart. While they worked, his wife drove around the country in the old company BMW trying to buy construction material for a new factory on the black market. By 1952, 61 people were employed at the factory in Grossbettlingen and a number of others in the company's numerous branch offices. Over the decade that followed, Herion continuously extended its range of products. When Fritz Müller joined the company in 1961, Herion was not only manufacturing valves, but also pneumatic and hydraulic control and regulation equipment for

Erich Herion, founder of Herion-Werke KG, where Fritz Müller worked in the design department between 1961 and 1963

automation processes in all areas of industry – from steel production and mining to elevator systems and water treatment plants. As the 'pacemaker of automation' (the company's advertising slogan), Herion exported products to countries as far away as the USA and Japan and helped drive the rationalisation of industrial process in the 1950s and early 1960s.

Erich Herion was not only a successful entrepreneur, he also looked after his staff and offered them a number of company benefits. 'The humanisation of social existence within a company, in other words the fostering of human relations', he wrote in 1962, 'is an important antipole to the rationalisation of industrial work.' He also realised at an early stage that 'mechanisation or automation renders the work of the unskilled labourer superfluous' and necessitates 'an increasing number of skilled and qualified labourers.' Moreover, he noted that 'the number of unskilled labourers is falling and the number of specialised workers and technicians is increasing; even the semi-skilled labourers are edging closer to the technicians.' Erich Herion reacted to this challenge by improving the training and qualification of his employees. Just like Bürkert in Ingelfingen, he attached great

The headquarters of valve manufacturer Herion in Fellbach near Stuttgart

importance to high-quality apprenticeships and preferred to recruit his apprentices from the surrounding region. He also tried to familiarise his young employees with the international activities of the company and encouraged them to continue training and educating themselves throughout their working lives.

Herion made a conscious effort to foster capable and motivated employees. Fritz Müller was one such an employee. His ingenuity and persistence soon caught the attention of Erich Ruchser, head of the development department at Herion. Ruchser (1919–93) was at the time one of the world's leading lights in valve development. He had given Herion more than 200 patents. The experienced development manager could see that Fritz Müller had the potential to succeed him in the company in the long run and recommended that Herion should start preparing the young designer for this managerial position. As part of this process, Müller was made head of the company's testing department.

Any designs to come out of the development department had to be tested in Müller's testing department. If improvements were needed, solutions were developed in conjunction with the development department. Both departments were so closely linked that the departmental heads worked hand in hand. The fact that he was now head of the testing department meant that Fritz Müller was free to develop his creative abilities in close co-operation with Erich Ruchser. At the same time, as the head of a highly qualified team, he gathered the leadership experience he would later need to run his own company. There was also an incredibly pretty female draughtsperson in the testing department. Müller quickly discovered that they could talk to each other about much more than just valve diagrams. 'It was easier to steal a quick

Erich Ruchser, head of development at Herion, recognised Fritz Müller's talent and arranged for his promotion to head of the testing department

kiss behind those huge drawing boards back then than it is now behind a PC screen,' recalls Müller with a grin.

However, things in his new department did not run smoothly for long. Through his close co-operation with Erich Ruchser, Fritz Müller had access to all the latest developments at Herion. Were Müller to accept a job offer from one of Herion's competitors, the damage to the company would have been considerable. This explains why shortly after his promotion, the management asked him to sign an amended version of his employment contract, which now included a non-competition clause. In signing, he would have agreed not to take on a comparable post at one of Herion's competitors or to establish his own company in the valve manufacturing sector within two years of leaving the company.

For Fritz Müller, signing the contract was out of the question. After all, he was still determined to go out on his own in the foreseeable future. Until that time, however, he wanted to continue gathering experience at Herion. For this reason, he attempted to renegotiate the clause to his advantage. He felt certain that Herion knew nothing of his plans to set himself up in business and was just afraid that he would return to Bürkert. Convinced that this was the case, he suggested that the non-competition clause be restricted to the state of Baden-Württemberg, which would have prevented him from returning to Bürkert. Müller reasoned that if Herion accepted his proposal, he would still be able to set up his own business outside Baden-Württemberg.

He was, however, wrong on all counts. Herion was outraged at the young designer's attempts to renegotiate the contract. Müller recalls the uproar: 'When I refused to sign the contract as it was, they started asking questions: did I know the opportunities this promotion opened up for me? How dared I make demands? And so on and so forth.' But Müller was resolute. Herion did not want to lose him and decided to let matters lie for the time being.

But the relationship between employer and employee had suffered a blow and Herion's mistrust had increased as a result of Müller's tactical manoeuvring. When a short time later, shortly before his 24th birthday, the young manager of the testing department bought a Mercedes 180 – the same model, incidentally, as driven by Erich Ruchser – and drove it to work one Monday morning, things came to a head. 'Although the Mercedes was an old, second-hand model for which I had been saving for a long time, the people at Herion saw the little star on the radiator grille and immediately asked themselves how a country boy from the Kocher Valley could afford a car like that.' Herion was convinced that this was the proof that Fritz Müller was selling development ideas to Bürkert. Müller was given an ultimatum: sign the contract with the non-competition clause or lose your job in the testing department. 'For me, the situation was clear,' says Müller, 'I told them: ok. I'll be on my way.'

In early 1963, only 18 months after he started work at Herion, he resigned. He was contractually bound to give Herion three months notice and Herion insisted that he fulfil his obligations to the company right up until the very last day. Naturally, he was no longer allowed to work in the testing department, but was instead given work filing documents in the commercial department. After eight excruciatingly boring weeks, Herion eventually released Fritz Müller from his contract. He was free. The time to set up his own business and manufacture his own valves had arrived.

CHAPTER 5

THE IDEA: A VALVE MADE OF PLASTIC

Fritz Müller's grandfather, Friedrich Müller, who lived in Esslingen, encouraged his grandson to realise his dream of setting himself up in business

Despite the fact that he had left Herion sooner than he had originally intended, Fritz Müller did not find his sudden independence a burden, but rather a welcome challenge. Even as a schoolboy making models, Müller knew that he had not only better ideas than the other boys, but also the technical skills and the necessary patience and determination to put his ideas into practice. He also knew that he had a rare talent. Fritz Müller did not see why others should benefit from all these characteristics. Going it alone was not about earning more money, but about being independent and creating opportunities to design. 'I didn't want to work somewhere for other people, I wanted to develop my abilities to the full,' recalls Fritz Müller. This would not have been possible as an employee or even as a manager. There was no way around it: he had to set himself up in business.

His grandfather, with whom he never lost contact down through the years, had always encouraged this yearning for independence. Every time Fritz Müller was in Esslingen his grandfather said: 'Fritz, see to it that you set yourself up in business. It doesn't matter if it's only a small workshop, just set yourself up in business.' But Fritz Müller wanted more than just a small workshop. After all, he wanted to own his own glider! He had calculated that in order to buy a glider, he would have to build up a company with at least 35 employees.

Fritz Müller was convinced that the only way to ensure his lasting success as an entrepreneur was to specialise. Having worked on valves at both Bürkert and Herion, it made sense to concentrate on valves. He didn't want to copy any existing models, but to develop completely new ideas. To do so, he didn't have to stay in Stuttgart. Just like his mother, he had never really felt at home in the city. As soon as he left Herion, he decided to return to Ingelfingen.

Before he moved away, he had something important to sort out, namely his relationship with Ingrid, the pretty draughtsperson at Herion. Although they had kissed behind the drawing board, they had never actually gone out on a date. Once Fritz Müller had been moved out of the testing department and given work filing documents for the last three weeks of his time at the company, they hardly saw each other. Then Ingrid took matters into her own hands and began visiting Fritz Müller at his new workplace every day. These meetings among the filing cabinets were the only happy times in Fritz Müller's dull final weeks at Herion. He was impressed by the fact that Ingrid was determined to stay true to him. He knew that he had found the woman with whom he wanted to spend the rest of his life.

There was only one problem: Ingrid was the daughter of Erich Ruchser, head of the development department at Herion. Given that Herion was already wary of Müller, it would only have made life more difficult not only for Fritz, but also for Ingrid and her father if the couple had met regularly. But Fritz Müller was not a man for secrecy and so, once he had left Herion, he asked Ingrid out to dinner. He told her that he would be leaving Stuttgart and would not be in touch with her for a few months. He promised that he would come back and asked her to trust him. Say what she might,

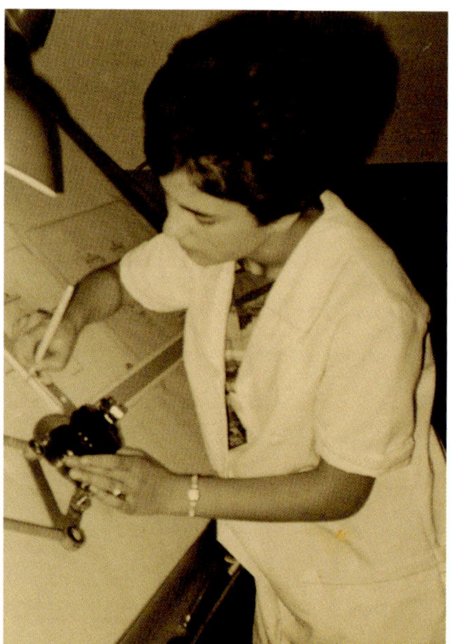

Ingrid Ruchser could not persuade Fritz Müller to tell her what he was planning. Nevertheless, she promised to wait for him.

Despite working hard in the testing department at Herion, Fritz Müller never lost sight of the important things in life

In Ingrid Ruchser, his boss's daughter and a technical draughtsperson at Herion, he found the woman with whom he wanted to spend the rest of his life

Fritz Müller now wanted to set himself up in business as quickly as possible. Only then could he return to Ingrid Ruchser without arousing suspicion that he was using her to spy on the development department at Herion. He moved back to Ingelfingen and spent weeks and months at his mother's house working on a valve design that would be different from any other valves that had been available on the market until then.

During his time at Bürkert, Fritz Müller had designed valves that he felt were ahead of their time. Later on, during his degree course, he had honed both his theoretical and practical skills. Moreover, the high standard of the development department at Herion had fed his ambition. 'I wanted to build valves that were better than those being designed at Herion at the time,' he recalls. One of his colleagues at Herion remembers that Fritz Müller was 'always coming around with some new idea for a new and improved valve.'

Müller might have shared some ideas with his colleagues, but he kept his best ideas to himself. In view of the fact that he had always had his sights set on working for himself, he began designing new valve models in the evenings after he had finished his day's work at Herion. He had made cross-sectional drawings and commissioned a small Stuttgart workshop to make prototypes of his designs at his personal expense. However, not having a test bench, he had to test his valves in his tiny attic flat in Stuttgart-Mettingen. During these tests, he discovered that some of his designs were less than perfect. 'I was very lucky in that the house in which my flat was located belonged to my grandfather,' says Müller, 'it is unlikely that any other landlord would have been so understanding about the fact that I kept accidentally flooding my flat during tests!'

At the time, Fritz Müller was looking for a technical alternative to the solenoid valve, which was used to shut off pipes. In a solenoid valve, a solenoid controls a tie-rod, which is used to open or close the pipe in question. The control current for the magnets can be switched on and off using any kind of switching element – be it a thermostat or a pressure-operated switch. Because of the speed with which solenoid valves can switch, they are used in many industrial sectors to control and monitor operational processes. Because of the broad range of applications in which they can be used and their low operating costs, solenoid valves are to this day in widespread use in the industrial sector.

However, companies at the time needed ever more powerful valves to cope with increasing pressures and pipe sizes (flow rates). This meant that although industries were moving towards smaller and more compact plants, more powerful and larger solenoids had to be used in valves. This problem was solved by using servo-controlled solenoid valves in which the magnetic coil was only used to relieve the main seal (diaphragm or piston), which was then lifted (actuated) and opened by the pressure of the medium flowing through (gas or fluid). This meant that relatively small solenoids could be used for valves that had to cope with high pressures and large nominal sizes.

Cross-section of a solenoid valve

In the early 1960s, these different types of solenoid valves

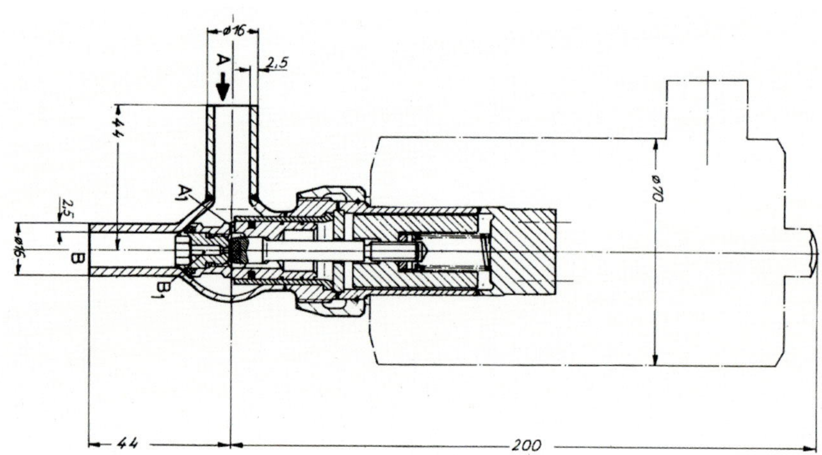

were already on the market, however, it looked as if they had reached their development limitations. In 1962, Herion launched its mini valve. It had a diameter of 3 cm and was 6 cm high. This was as small as this kind of valve could get; any smaller and the switching frequency would cause the solenoid to burn out. The only way to get around this problem was to use pneumatic valves, i.e. valves that were actuated using compressed air. However, only very few industrial companies had the compressed air systems needed to run such pneumatic valves. The challenge was, therefore, to develop a highly compact electric valve that could develop forces similar to those generated by a pneumatic valve. This is what Fritz Müller set about doing in his attic flat in Stuttgart-Mettingen. 'I wanted to move away from solenoids and pneumatics,' he recalls. In order to do so, he made use of design principles that had been used in cooling water thermostats as far back as the 1920s and tried to adapt them for use in a new valve control mechanism. He spent many nights painstakingly developing a valve in which the sealing diaphragm was no longer actuated by a solenoid-controlled tie rod, but by a cylindrical metal bellows filled with a highly volatile substance. When this medium was heated using an electric heating coil, it expanded,

Functional description and cross-sectional view of a servo-controlled solenoid valve

Funktionsbeschreibung

In Ruhestellung (bei nicht erregtem Magnet am Steuerventil) ist das Ventil geöffnet.
Bei Erregung des Magnets (13) am Steuerventil (14) schließt das Ventil.

Schließvorgang
Bei Erregung des Magnets (13) am Steuerventil (14) öffnet der Vorsteuersitz (15), Vorsteuersitz (16) schließt, der Membranantrieb (17) wird mit dem Steuermediumdruck Z beaufschlagt, die Spindel (18) bewegt sich entgegen der Kraftwirkung der Feder (10), wodurch das Ventil schließt.

Öffnungsvorgang
Bei Entregung des Magnets (13) am Steuerventil (14) öffnet der Vorsteuersitz (16), Vorsteuersitz (15) schließt, der Steuermediumdruck Z am Membranantrieb (17) baut sich über das Steuerventil (14) in den Rücklauf ab; durch die Kraftwirkung der Feder (10) hebt die Membran (11) vom Ventilsitz ab und öffnet das Ventil.

Die Durchflußrichtung des Mediums ist beliebig.

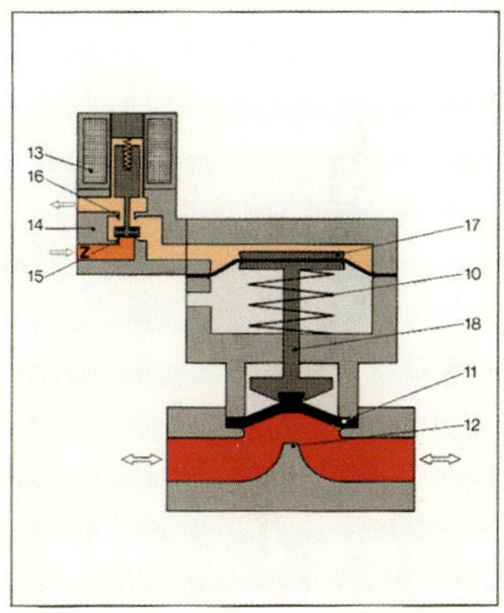

Water treatment plant in Rotterdam, late 1960s: thanks to pneumatic and electronic control and regulation equipment, this plant, which had a capacity of 130 million litres of water per day, was operated by a team of only five members of staff

which in turn caused the bellows to expand and press the diaphragm downwards, closing the valve. When the heating coil was switched off, the medium in the bellows cooled down, contracted, relieving the load on the diaphragm, which then opened under the pressure of the medium flowing through the body of the valve. If Fritz Müller's calculations were correct, this valve could be more powerful and more resistant than a comparable solenoid valve.

That was the theory, anyway. Fritz Müller got a prototype made and tested it at home. The metal bellows developed a much greater force than a solenoid controlled tie-rod of the same size. However, it built up this force relatively slowly, namely at the speed at which the medium inside the valve heated up and expanded or cooled down and contracted. This meant that Fritz Müller's valve reacted more slowly than magnetic or pneumatic valves. Changing the medium in the valve did not make any difference. In short, the valve was not suitable for rapid, automated work processes. Having worked on the valve in his attic flat for several months, Fritz Müller had to accept that his attempts to design a completely new drive principle for industrial valves had failed.

But Fritz Müller was not to be put off by this setback. He instead turned his attention to another problem faced by valve manufacturers in the early 1960s. 'It was around about this time that protection of the environment became an issue for the first time,' says Müller. 'Strong acids and alkalis are a dominant feature of applications in the field of electroplating in particular. Up until then, companies had just pumped their waste water into the nearest stream. In the early 1960s, they were suddenly obliged to install water purification and neutralisation equipment.'

The valves that were used in this equipment – even those that were made of expensive, high-quality stainless steel – were quickly corroded by the aggressive media with which they came in contact. All the major valve manufacturers were searching for a solution to this problem and had not, as yet, come up with one. Müller remembers: 'By this time,

pipelines made of acid- and alkali-resistant plastic were already being used for electroplating, and the first manually operated valves made of plastic were also available. However, these manually operated valves were not accurate enough to be used in industrial applications. Industry was crying out for solenoid valves. When it came to such automatic industrial valves, valve manufacturers needed heat- and pressure-resistant materials and therefore used only those materials that were known to be of a high quality – stainless steel, for example. Valve manufacturers did not at the time consider PVC to be a suitable material for valves; it was at most used to protect the metal parts in valves against pitting. Herion had, for example, attempted to coat the tie-rod and body of solenoid valves with plastic so that they could be used in neutralisation plants. The problem was that this coating made the solenoids less effective. They just couldn't make any progress in this regard.'

It was Fritz Müller who made the breakthrough shortly after leaving Herion in the spring of 1963. 'I thought that if the pipes could be made of plastic, then I could make the valve bodies out of plastic too,' he recalls. And so, the idea of the plastic industrial valve was born. It is possible that Fritz Müller was able to make this small, yet decisive step because he was working on his own. No longer influenced by the prejudices of his colleagues, he was able to free himself from old ideas. It is also possible that he was unconsciously influenced by a technical material revolution in a completely different sector, namely the construction of gliders. In the late 1950s, glider manufacturers began experimenting with glass fibre-reinforced plastic. This material was not only lighter than the wood and aluminium that had been used up until then, it also boasted vastly improved aerodynamic properties. Even before Fritz Müller joined Herion in Stuttgart, the Institute for Aircraft Construction at the University of Stuttgart and the local gliding club had manufactured the very first glider made of glass fibre-reinforced plastic. In the years that followed, aircraft manufacturers around the world refined the construction method.

The young Fritz Müller dreamed of owning one of Björn Stender's legendary BS 1 gliders

Fritz Müller, who had continued gliding during his eighteen months at Herion, had followed these developments with interest. In late 1962, the BS 1 glider, made by the young aircraft designer Björn Stender (1935–63) caused a stir in the world of gliding. This high-performance glider, which was made of plastic, reached maximum speeds of up to 300 km per hour and glided almost noiselessly through the air. It had a maximum glide ratio – i.e. the ratio between loss of altitude and distance covered (glide path) – of 1:44, more than twice that of a conventional glider. These outstanding flight characteristics made the BS 1 the most elegant and powerful glider of its day. When Björn Stender set off on Whit Sunday 1963 from the glider airfield in Stuttgart-Hahnweide on a long-distance flight to Fontainebleau near Paris (a distance of over 500 km), Fritz Müller was one of a number of gliding fanatics from the region who came to take a closer look at Stender's 'Super Orchid'. Fritz Müller was so excited by what he saw that he began to make concrete

plans to realise his long-held dream of owning his own glider. 'I wanted a glider just like that. I wanted a BS 1,' says Müller, looking back on the Stuttgart Air Show.

It could well be that the elegant glass fibre glider designed by Björn Stender, who was only a few years older than Fritz Müller, inspired Müller to use plastic in his valves – a material rejected by other designers. Whether this was the case or not, Müller spent the summer of 1963 working hard on his idea of a plastic solenoid valve. However, several obstacles had to be overcome first. Even if he managed to make a valve body out of plastic, he would still have to protect the tie rod, which was made of a ferrous metal, and the solenoid that he would need to operate the valve against corrosion by aggressive media. As already illustrated by tests at Herion, coating these metal components with plastic was not a satisfactory solution. Eventually, Fritz Müller hit upon the idea of hermetically sealing the tie-rod and solenoid from the valve body using a thin, yet highly elastic elastomer diaphragm, thereby preventing it from making contact with the aggressive fluids. However, in initial tests, the diaphragm was pushed up and extended to such an extent by the pressurised medium that was flowing through the valve body that it either quickly tore or dissolved. Fritz Müller initially tried relieving the pressure on the diaphragm by using a stronger tie-rod spring, but this would have necessitated a relatively large spring and a correspondingly large magnet. This was not compatible with making the most compact valve possible. He eventually came up with the idea of filling the empty space behind the diaphragm with highly viscous silicone oil. This hydraulic counter pressure relieved the pressure on the diaphragm and meant that a diaphragm made of much thinner rubber could be used, thereby keeping down the size of both the tie-rod spring and the solenoid.

Having worked away on his own in a flat in the house owned by his mother in Ingelfingen, Fritz Müller finally brought his drawings to the precision engineering workshop and asked for a prototype of his valve to be made. Although the finished valve was 12.5 cm high and had a

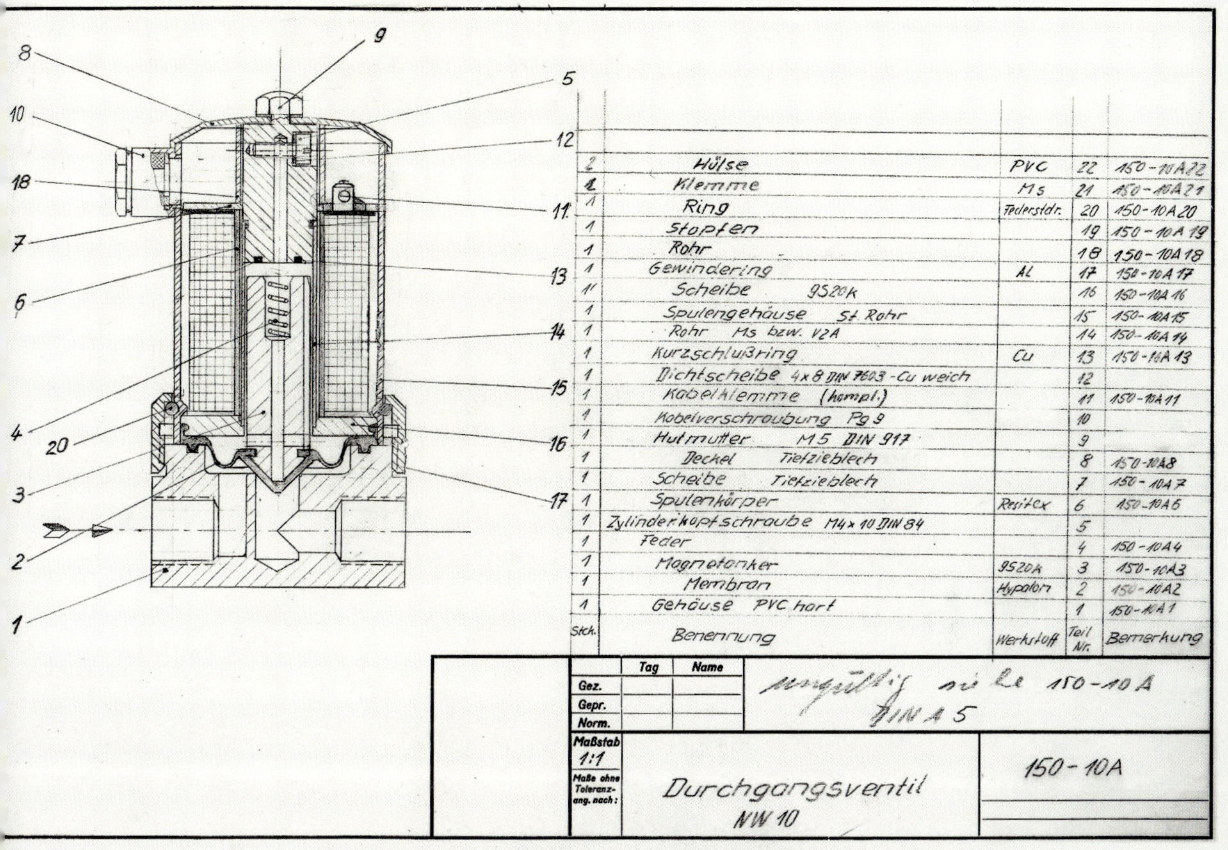

diameter of 6 cm – making it almost twice the size of Herion's mini valve – it was still very compact for a valve with a nominal width of 10 mm. The practical tests run in the weeks that followed showed that the valve was indeed able to withstand aggressive media. The PVC valve body also proved to be extremely acid- and alkali-resistant and the pressure-relieved elastomer diaphragm protected the tie-rod and the solenoid perfectly even under extreme loads.

Fritz Müller's cross-sectional diagram of his very first valve (1963)

The prototype of Fritz Müller's first valve (1963)

Fritz Müller had managed a breakthrough! He had developed a completely new valve that exactly met the requirements of the market and was also less expensive to manufacture than the competition's less robust products. He knew he could now set himself up in business. With his commercial training, his elder brother Kurt would have been an ideal business partner, but the young father was not willing to take what seemed to him to be the risk of leaving his secure and well-paid job in Künzelsau to join forces with his brother. Nevertheless, the fact that his brother did not join the company did not prevent Fritz Müller from calling his new company Gebrüder Müller Apparatebau (Müller Brothers Plant Engineering), or GEMÜ for short. 'I was still very young and I thought that introducing myself as "Fritz Müller of Müller Brothers" rather than just plain "Fritz Müller" would earn me additional respect at business meetings,' says Fritz Müller with a grin. On 11 November 1963, Müller registered his company in the town of Ingelfingen. As soon as he had, he picked up the phone, dialled Herion's number, and asked to be put through to Ingrid Ruchser.

CHAPTER 6

HOME-MADE VALVES: THE EARLY YEARS (1963–65)

Despite the fact that after their last dinner date, she had not heard anything from him for more than six months and had no idea where he was or what he was doing, Ingrid Ruchser kept her promise and waited for Fritz Müller. After he called her at the office in late autumn 1963, they began seeing each other on a regular basis. Naturally, Fritz visited Ingrid at home. He got on very well with her father, Erich Ruchser. Both men were inventive and loved designing and constructing things. Nevertheless, relations between the two were not easy. Because of his contact with Fritz Müller, Herion began to question Erich Ruchser's loyalty to the company.

To this day, his daughter is convinced that Erich Ruchser had the ingenuity and skill to set himself up in business as a self-employed engineer. However, without the necessary commercial ambition for such a move, he preferred to continue working as a senior employee at Herion. In 1952, Erich Herion had personally hired Erich Ruchser as plant manager in Fellbach when he heard that the highly talented young engineer and father of two small children had suddenly been laid off when his employer had gone bankrupt. From that day on, Erich Ruchser's loyalty to Herion was unshakeable.

Fritz Müller and Ingrid Ruchser on their wedding day at the end of August 1964: unhappy with the old-fashioned hair style created by a local hairdresser on the morning of her wedding, the young bride rushed home, redid her hair herself, and put on a veil for the ceremony

This is why his colleagues' jibes that he must surely be talking business with Fritz Müller when he came to visit hurt him so deeply. Erich Ruchser hoped in vain that the relationship

between his daughter and Fritz Müller would not last long and that it would, with a little assistance from himself, peter out. On the contrary, 18-year-old Ingrid told her father straight out that she didn't care what he and his colleagues at Herion thought of the relationship; she was going to stand by Fritz Müller. 'It was an impossible situation for Erich Ruchser,' recalls Fritz Müller. 'He wanted to be fair to his daughter and he wanted to behave as decently as he could to me ... and all the while he had to put up with his colleagues' snide remarks about the situation.'

It was only after the couple got married in Ingelfingen on 29 August 1964 that the unpleasant situation improved a little. Ingrid Müller remembers: 'We got married in the registry office on the Friday morning, celebrated with our friends on the Friday evening, got married in church on the Saturday, rested on the Sunday, and got up early on the Monday morning. We didn't get up early to set off on a romantic honeymoon, but to set to work.' The young entrepreneur urgently needed help with his accounts. 'After breakfast, my newly wed husband put a cardboard box on the table. It contained all the receipts and bills that had accumulated since he had registered his business ten months previously. I was supposed to sort out the receipts needed for the Tax Office. That was how I started out at GEMÜ.'

It was certainly not an easy start for the young bride. As a draughtsperson, Ingrid Müller knew very little about the commercial side of business. In those first few months, she regularly had to pay interest on arrears to either the Tax Office or the health insurance company because she had – either out of ignorance or an excess of work – neglected to make the necessary payments. However, with the help of her mother, a trained bookkeeper, who quickly came from Stuttgart to help out her daughter, Ingrid Müller soon learned the basics of keeping a cashbook and wage accounting. Soon after, she was able to rely on the help of an accountant.

While Ingrid Müller was busy learning how to keep the back office up and running and typing letters and invoices on an ancient typewriter, her husband was busy building up his company. True to his motto that 'an idea is only good if you make something of it', he quickly began making preparations for production as soon as his plastic diaphragm valve prototypes and a plastic angle seat globe valve of a similar design had successfully completed an endurance test.

To get the ball rolling he needed start-up capital, which meant getting a loan from the bank. He himself had no securities to offer, but his mother, who was firmly convinced that her son's enterprise would be a success, furnished the necessary securities. As the daughter of Ingelfingen's former cellar master, Marie Müller was well-respected in the town and on the basis of her solid reputation, the bank was willing to lend Fritz Müller DM 5,000.

Fritz Müller, who had up until that point paid for everything out of his own savings, was not happy about having such a large debt to repay. It didn't help that an older master craftsman he knew told him that he would soon get used to handling much larger loans. In order to keep his debts to a minimum, Fritz Müller used his start-up capital very sparingly. He did not rent any premises for his

GEMÜ's first premises: Marie Müller's house in Ingelfingen

first workshop, but instead set up shop in his mother's garage on Hundsbergweg in Ingelfingen. When it came to equipment, Fritz Müller only bought what was absolutely necessary. In the early days he made do with a hacksaw and a lathe. He bought his first lathe from a company that wanted to sell the machine for 86 pfennigs per kilo. In order to keep the price down, Müller removed as many parts as he possibly could on the way to the scales.

Getting the necessary materials was also problematic. In the days when orders were few and far between, Fritz Müller only needed small amounts of material. But small amounts were not easy to come by. For example, he could only buy the small amounts of steel he needed from a large steel wholesaler in Stuttgart. 'They could spot me and my rusty old Volkswagen at a distance,' says Müller, 'and whenever I drove over there from the Kocher Valley, I could almost hear them thinking: "here he comes again, looking for a few miserable rods of steel!"'

Bought second-hand and cannibalised to provide material for a valve: GEMÜ's first work bench

In view of the fact that Fritz Müller calculated the material he needed very carefully, it was essential that no mistakes were made. Sometimes, however, things went wrong, as they did one Sunday afternoon in the first year of his business. Müller had been working flat out the entire day under great time pressure. He was preparing a valve to present to a potential customer the next morning. In his haste, he lost concentration for a moment and damaged the solenoid armature. The damage could not be repaired and Müller knew that he would have to start again from scratch. The only problem was that he had run out of steel. There was no point even trying to buy steel in the Hohenlohe region on a Sunday afternoon in the 1960s and he knew he did not have enough time to drive to Stuttgart. The innovative engineer then decided to saw up the tracks on which the lamps on his lathe were mounted and used the steel to make a new solenoid armature. The next morning, he was able to present the customer with an impeccable valve.

However, even more so than was the case with steel, getting the necessary small amounts of PVC for his valve bodies proved very difficult. 'Initially, I needed only about a half a metre or, at best, maybe about 80 cm of PVC and when I asked a plastic dealership, I got the same reply: "No such thing as a half metre; only half tons or tons."'

That being said, Fritz Müller soon realised that lathing and milling valve bodies from solid blocks of PVC was much too time- and material-consuming. Production had to be economical. He began searching for a supplier that could manufacture his valve bodies using injection moulding techniques. In the late summer of 1964 he found a co-operation partner in the pipe manufacturer Georg Fischer of Schaffhausen, who had been manufacturing PVC pipes since 1957. Fischer used Fritz Müller's drawings to make injection-moulded PVC valve bodies and also included the finished diaphragm valves in his range of products. Fritz Müller had his first customer.

It was the start of better things to come: through Fischer, Müller made contact with an Italian PVC manufacturer in Genoa. Soon after, when this manufacturer's German representative, Mühlheimer Kunststoff, exhibited at K, the international plastics trade fair in Düsseldorf, Fritz Müller was invited to exhibit his valves at the company's trade fair stand. 'Naturally, it wasn't my own trade fair stand,' recalls Müller, 'I just had a brightly painted wooden board on a steel tube stand. I displayed my two valve models – my Type 202 diaphragm valve and my first angle seat globe valve – on this board alongside a brochure that I had designed myself. It was all very simple, but very well made. Even back then, I attached great importance to the appearance of the valves and the way they were displayed.' Good design had always been important to Fritz Müller. He knew that being new on the market, he only had a chance of succeeding if his products were not only innovative and inexpensive, but also functioned perfectly and looked good too. He knew that appearance was the first step towards earning the trust of the customer. 'Technology is beautiful and many people working in technical professions have an eye for this kind of beauty,' he explains. 'I wanted these people to look at my valves and immediately think: "that would look good in my plant."'

Even so, as a new kid on the block, it was not easy for Fritz Müller to attract attention among all the established valve manufacturers. Moreover, many trade fair visitors found his valves too small and simply didn't believe his claims regarding their performance. Only one manufacturer of neutralisation plants ordered five valves so that he could test them – more out of curiosity than any firm conviction that they would be worth the investment. His contact with Herion, on the other hand, was much more promising. It goes without saying that representatives of his former employer came and took a closer look at his display.

GEMÜ's first trade fair panel (1964)

They immediately realised that Müller's plastic valves meant tough competition for Herion's products. 'That really set the cat among the pigeons!' recalls Müller. It was not long until the head of Herion paid him a visit and made him a surprising offer: 'Erich Herion came to see me and said: "Mr Müller, what do you say? We'll invest DM 40,000, you can add your material orders to ours, and when we no longer need old machinery, you can have it free of charge. We'll include your valves in our catalogue, customers can order your valves through us, and we'll buy as many valves as necessary."' For Müller, it was as if all his problems had been solved at once. He could repay his loans, did not need to worry about material procurement or new machinery, and Herion would assume responsibility for sales. He accepted Erich Herion's offer without hesitation.

Initial orders soon started coming in and Fritz Müller was at last in a position to hire staff for production. He found his very first employee during a visit to his local barbershop in late September 1964. Sitting in the next seat was Günter Nägele, whom he knew from their days at school together. They got chatting and Fritz Müller told him about his company. Günter Nägele asked if he could do with a helping hand. When Müller said that he could, Günter Nägele didn't waste any time. He handed in his notice and finished work in nearby Weissbach on 30 September 1964. At 7.00 a.m. the next morning he knocked on Müller's front door. To a surprised Ingrid Müller he announced: 'Nägele. Günter. I'll be working here from now on.' And he did. Wherever he was needed, Günter Nägele rolled up his sleeves and lent a hand.

Others quickly joined the team. In order to ensure that production ran smoothly even when he was out on business, Fritz Müller hired Wilhelm Arnold as his works manager. The two knew each other from vocational school. In the months that followed, GEMÜ began to occupy more and more rooms in the house on Hundsbergweg. In the garage, Günter Nägele used an old screw press to press rubber diaphragms or solder valve sleeves. In the attic, two women wound wires for the solenoids. In Marie Müller's kitchen, the solenoids that were covered in epoxy resin were hardened in the oven, while the valve sleeves were cleaned in large saucepans on the cooker to remove any soldering residue. After that, the various parts were assembled by another woman. The finished valves were then checked, packed, and brought to the post office in Ingelfingen by Ingrid Müller. 'If something wasn't quite right on the delivery note, the people at the post office never tired of telling me how much better things worked at the other companies in Ingelfingen,' recalls Ingrid Müller. 'The fact that these companies were much bigger, had their own sales and dispatch departments, and at least 20 years more experience than us, didn't matter a jot to them.'

Despite these teething problems, business quickly picked up. It soon became clear that the old machines were no longer suitable for large-scale production. Fritz Müller needed special machinery and tools that were exactly suited to his requirements – the kind of equipment he could only get from a good toolmaker. He found his man in Helmut Weber. Weber and Müller had been at primary school together and had started their apprenticeship at Bürkert at the same time, the only difference being that Weber was training to be a toolmaker and not a precision engineer like Müller. After that they lost touch. While Fritz Müller was studying, Helmut Weber worked for a company in the city of Heilbronn. He later moved to Bosch and returned to Ingelfingen once he had completed his military service in 1961. In the spring of 1965 he heard from his friend Günter Nägele that Fritz Müller was looking for a toolmaker who could make special machinery and tools for the growing company. Weber did not feel challenged by the job he had at the time and was very interested in the job at GEMÜ. He also, as he recalls, 'found it more exciting to be involved in the start-up of a new company.' 'I knew that I would be able to contribute my ideas at Müller's company. So I packed in my old job and started working for Fritz.'

The working conditions for Helmut Weber in those early years were anything but ideal. When the garage at the house on Hundsbergweg got too small, Fritz Müller rented a second garage in the neighbourhood and turned it into a makeshift workshop. 'The metal door was lined with polystyrene so that it would not get too cold in winter. And apart from the lathe, there wasn't much else in there,' say Helmut Weber and Günter Nägele. Weber now spent Monday to Friday working in this garage on the lathe. At the weekends, he and Fritz Müller set to work designing and making new machinery. Their first project was to come up with a new rubber press for the diaphragms. Müller could only produce 50 diaphragms per week using the old screw press; he needed between 150 and 200. 'It took us two weeks to get our self-made press up and running smoothly. During that time, we were unable to produce anything,' recalls Müller. 'Naturally, the customers protested and I had to field a few very unpleasant phone calls, but when the new machine went into operation we were suddenly able to produce five times as quickly as we had been before. Our turnover increased accordingly.'

GEMÜ was in the black and doing well. Nevertheless, this did not mean more money for Fritz Müller and his family; the profits were channelled straight back into the company. New machines were bought and constructed and loans repaid. 'In the first few years, I allocated a monthly housekeeping allowance of exactly DM 50 to the family,' recalls Ingrid Müller. 'Whatever else we earned went straight back into the company.' Even the generous financial gift that Erich Ruchser had given the young couple as a wedding present was immediately invested in new machinery.

As the company's turnover grew, so too did the working hours. '50 to 60 hours a week was the norm; more when things were busy,' says Helmut Weber. Fritz Müller frequently burned the midnight oil in his design office. Ingrid Müller remembers: 'I often spent my evenings and, if needed be, also the weekends in my tiny office, doing the books and dealing with correspondence. I had just turned 20; it was sometimes tough, especially in the summer. But my husband and I – all of us, as a matter of fact – rose to the challenge of having to keep going, and to work harder, faster, and better.'

Marie Müller, seen here with Fritz Müller's brother, Kurt, was the rock of the family in the early years

Fritz Müller with his daughter Regina (1965)

Amidst all the hustle and bustle, Marie Müller was a supportive and a calming influence on everyone. Soon after Ingrid and Fritz Müller's wedding, she gave up her bedroom and slept on the couch in the living room. After the birth of her granddaughter Regina in 1965, she helped her daughter-in-law look after the child. If necessary, she was also willing to help wind coils and keep an eye on the hardening of the valve coils in her oven. She also cooked for the family and, on one occasion, unintentionally for the staff too! Günter Nägele remembers: 'One day, I was in the kitchen cleaning the solenoid sleeves. There was a big pan of home-made hamburgers on the cooker for the two Müller brothers. Marie Müller stepped out of the kitchen for a moment and I took the pan and distributed the hamburgers among the staff in the house.' Helmut Weber continues the story: 'Of course, we were all delighted to get the hamburgers. It was only when we heard the uproar upstairs – poor Marie Müller didn't know whether to laugh or cry – that we realised that Günter had not asked the head cook's permission before making off with the hamburgers.'

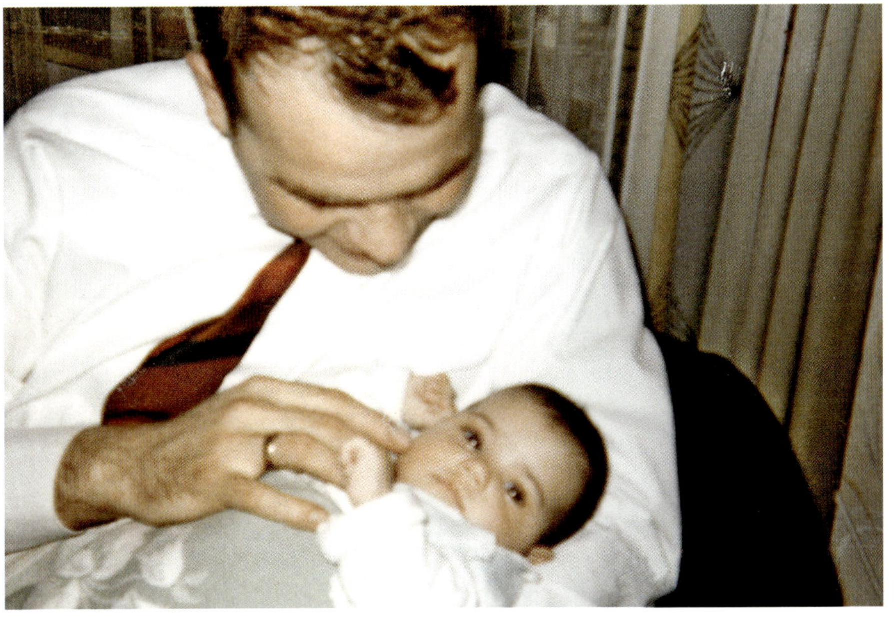

CHAPTER 7

EARLY SUCCESSES: GEMÜ'S GROWTH AND DEVELOPMENT (1965–68)

GEMÜ was one big family that continued to grow and grow – a fact highlighted by Günter Nägele's distribution of Marie Müller's homemade hamburgers. Ingrid Müller had her hands full with the accounts and the back office, leaving her no time to work as a draughtsperson. Her husband, Fritz, needed help in this regard and so, shortly after Helmut Weber joined the company, a draughtsman in the shape of Dieter Specht was hired. More workers were soon needed to complete the ever-increasing number of incoming orders on time and the workforce grew accordingly.

GEMÜ was receiving regular orders from Herion and Georg Fischer in Schaffhausen. However, other customers that Fritz Müller had acquired himself were gaining in importance. One such company was the Neuss-based Dr. A. Kuntze GmbH, whose boss had bought his first five GEMÜ valves from Fritz Müller at the plastics trade fair, K, in Düsseldorf. Kuntze manufactured electroplating equipment in which aggressive acids and alkalis had to be transported. The company had tested GEMÜ's plastic valves and was so impressed by their performance that it was now ordering 50 units a week. True to his motto that a company should treat every customer as if it is its only customer, Fritz Müller often assumed responsibility for the dispatch of the valves himself. 'Every Friday, I took the box of 50 valves to the train on my bike,' he recalls. 'Sometimes I was so late that I had to pedal after the train to ensure that my valves would get off on time.'

Unfortunately, the system with which GEMÜ planned to make a big impact on potential customers was only operational on the last days of the Hanover Fair in 1965

Had he not exhibited at the plastics trade fair, Fritz Müller would never have made contact with Kuntze. He had learned while working at Bürkert and Herion how important trade fairs were for customer acquisition and decided to exhibit at the Hanover Fair in 1965. Naturally, the young entrepreneur could only afford a tiny stand. He decided that in order to make up for the small size of his stand, he would need an impressive exhibit. Because visitors to the plastics trade fair had expressed doubts about his claims regarding the valves' performance, Fritz Müller wanted to show his valves in action at the Hanover Fair. He designed and built a complex system of transparent pipes in which a variety of GEMÜ valves controlled the flow of water. Helmut Weber remembers: 'Fritz had been working until 5 o'clock in the morning on the exhibit. Two hours later, we packed it into the trailer and set off for Hanover in the rickety old Ford Taunus. I drove; Ingrid Müller sat on the back seat behind me; Fritz was beside her – out for the count – his legs dangling over the back of the front passenger seat.'

The first of many patents: the letters patent for Fritz Müller's plastic solenoid valve (1965)

Once in Hanover, Fritz Müller and Helmut Weber spent day and night working on the demonstration model. The system had worked perfectly in a test run back home, but on the first day of the fair, it inexplicably began to leak. Müller and Weber spent hours searching for the cause. Eventually, with heavy hearts, they had to empty the water out of the model, knowing that they could not now show potential customers the valves at work. It was only during one of the sleepless nights that followed that the two succeeded in repairing the system, which went into operation on the final days of the fair. 'Nevertheless, it was still a huge setback,' recalls Helmut Weber.

This technical hitch was not the only low point of their visit to Hanover. While Weber and the Müllers were trying hard to convince trade fair visitors of the high performance of GEMÜ valves (albeit without the demonstration model), Christian Bürkert suddenly appeared at their stand. Müller remembers: 'He took one look at my valves and was virtually speechless with rage. He eventually managed to say that he simply would not tolerate the fact that I, a former apprentice in his company, dared to make valves. "There is only one person in the Kocher Valley who makes valves," he thundered, "and that is me: Christian Bürkert!" He went on to say that I, Fritz Müller, need not think I was going to compete with him. He would just not stand for it! He was going to start making valves like mine and then we'd see where I'd be.'

But Fritz Müller was not about to be unnerved by his former boss's outburst. He knew that the patent application process for his plastic valve was nearing completion and that Bürkert did not at the time have any plastic valves for aggressive media on the market. Müller remembers: 'That was the way things were in the Kocher Valley back then: the larger companies tried to keep the smaller companies down. But quite apart from that, Christian Bürkert had no reason to be furious with me. He could simply have acknowledged the fact that I had gone my own way. He could have said to himself, "Fritz Müller learned his trade in my company, came up with a good idea, and is now making completely

BUNDESREPUBLIK DEUTSCHLAND

Urkunde

über die Erteilung des Patents

1 227 299

Für die in der angefügten Patentschrift dargestellte Erfindung ist in dem gesetzlich vorgeschriebenen Verfahren

dem Herrn Fritz Müller, Ingelfingen

ein Patent erteilt worden, das in der Rolle die oben angegebene Nummer erhalten hat. Das Patent führt die Bezeichnung

Magnetventil

und hat angefangen am 29. August 1965.

Deutsches Patentamt

Pat.- Rolle 3a

different valves to me." I may have remained calm on the outside, but Bürkert's performance left me feeling very uneasy. This was in fact the start of the rivalry between us,' says Müller looking back.

'I refused to be intimidated by this incident and spent the remaining days of the trade fair making contact with as many potential customers as possible. I wrote a report about everyone who came to the stand. Back home in Ingelfingen, I assessed the reports, wrote to the people I had met, and made appointments to show them my valves,' recalls the entrepreneur. It was worth the effort. GEMÜ received a flood of new orders after the Hanover Fair. 'In 1964, I posted DM 14,600 turnover,' says Müller, 'in 1965 my turnover was more than twice that.'

A lot of effort had gone into earning this money. Many potential customers were wary of Müller; he seemed too young to be a reliable manufacturer. He was once asked by a customer whether he was 'the son of the old Müllers'. When Müller replied that he was, the customer was reassured and did actually show an interest in GEMÜ valves.

Other encounters did not go as smoothly. 'In late 1965 I began replacing the selenium rectifiers in my valves with silicone rectifiers that performed better,' says Müller. 'A short time later, an engineer working for an important customer in the Münsterland region called up, wanting to know what was wrong with my valves all of a sudden: they kept on burning out. I began to feel queasy on hearing this because I immediately suspected that the new rectifiers were causing the problem.' Because the problem could not be solved on the phone, Fritz Müller got into his dented Ford at 4 o'clock the next morning and drove to the customer's premises. 'Once there, it quickly became apparent that the problem was not the rectifiers themselves, but the fact that the customer had installed the valves directly alongside a number of large engines. These engines were switched on and off at regular intervals. Switching off the engines generated an induction voltage that could peak at well over 2,000 V. No rectifier on earth could withstand that!' Fritz Müller and the engineer resolved the problem by installing a surge protector device. 'The engineer was thrilled with the rapid service response. I was exhausted and just wanted to go home. But while I was on my way to the car park, one of the company secretaries told me that the boss wanted to see me. I thought he was going to congratulate me on a job well done. How wrong I was! Enthroned in his office behind a huge dark desk, he wanted to know how many people I employed. Glancing around his oak-panelled office, I saw an oil painting of a stately bearded gentleman, the grandfather of the company's founder, and numerous awards that the company had won since eighteen-hundred-and-god-knows-what. I knew that I would be out on my ear if I told him the truth. So I fibbed and said "15". To which he replied: "I cannot do business with a company as small as yours. You have a long journey ahead of you and I have much to do. Goodbye."' Shocked and numbed by the encounter, Müller made his way to the car park, watched from his office window by the company boss. 'He wanted to see what kind of an old jalopy I was driving,' says Müller recalling the scene. Then, to crown it all, the car wouldn't start. Müller had to hotwire the old Ford before he could drive away. 'I have no memory of the journey home that day. But when I got back to Ingelfingen, the first thing I did was to instruct the team never to send valves to that company again.'

There were, however, other entrepreneurs who valued the ingenuity and flexibility of their young colleague in Ingelfingen. At the plastics trade fair, K, Müller had made contact with the Swiss family-run company Bachofen, which was the industrial agent for a number of large Swiss companies in Europe. Initially, Müller heard nothing from the Swiss after the Hanover Fair. Then one day out of the blue, the two Bachofen brothers contacted him and announced they would soon be paying him a visit in Ingelfingen. 'The next day, they drove up to the house on Hundsbergweg in their Jaguar XS and asked for a tour of the plant. There was no point spinning them any yarns, so I asked them to step into my tiny little office with its second-hand furniture. I pushed two chairs up to the table and opened a bottle of my best white wine. They were delighted! "At last," they said, "a decent morning pick-me-up and not just another cup of

The Müllers moved into a flat in this house in Niedernhall after the birth of their daughter Regina, whose bedroom served as GEMÜ's design office during the day

coffee." Then I took them on a tour of my "plant". They saw the attic, the kitchen, and the garage. I also showed them my first lathe and explained why the track on which the lamps should have been mounted was missing. They must have been impressed by some part of my story because from that day on, they included GEMÜ products in their range.'

Additional orders came in through Bachofen, making it necessary to hire more people. GEMÜ's very first Christmas party in December 1965 was attended by ten employees. Despite the fact that Fritz Müller had rented another garage as a workshop and had moved with his wife and daughter into their very first home, a three-roomed flat in the neighbouring town of Niedernhall, the company was soon too big for the house on Hundsbergweg. As Ingrid Müller explains, the family flat soon became a branch of GEMÜ. 'The child's bedroom served a number of functions: at night, our daughter Regina slept there, from 7.00 a.m. to 12.00 midday and from 1.00 p.m. to 4.00 p.m. it served as the GEMÜ design office, and in between, we used it as a dining room.'

Under these conditions, there was hardly any room for a real home life. Fritz Müller soon realised that he would not be able to continue running his company from the family kitchen, living room, and garage for much longer. Early in 1966, he began looking for a 1,000-1,500-square metre site that would be suitable for a small production facility. Whether it was that other local companies used their influence on the mayor to block his requests or whether it was that Ingelfingen, a rural community, was simply not prepared for the arrival of new industry, all Müller's requests to the town of Ingelfingen regarding a site for his factory were turned down.

He had more success in Criesbach, a small community only 1.5 km away from Ingelfingen. The only industrial company in Criesbach, a small bitumen factory that had supplied the material needed for the construction of the Kocher Valley highway, had closed down in the late 1950s. Because the premises of this company at Criesbach railway station were not in use, the mayor of the town, Otto Betzold, made the site available to Fritz Müller on a hereditary lease in 1966.

The GEMÜ production facility (right) was situated directly beside Criesbach railway station (left)

As early as 1968, the new production facility had to be extended – not least in order to house also Fritz Müller's first glider

In May 1966, the staff of GEMÜ moved into the prefabricated production facility they had erected themselves at Criesbach railway station. It quickly became apparent that the facility was too small. It was decided that the hall should be extended. Initially Fritz Müller intended to oversee the construction of the extension himself. Then, however, another childhood friend, Emil Fick, got in touch with him. 'Emil Fick was very good with his hands,' says an admiring Fritz Müller. 'No matter what you asked him to do, he always did the job well.' Müller knew, therefore, that he could quite happily leave Emil Fick to organise and manage the extension project. Fick remembers: 'When I applied to Fritz Müller for a job back then, he told me straight out: "If you want to work for me, you'll have to build your workplace first." And so I was hired as overseer.' Working outside office hours under Emil Fick's leadership, a group of labourers spent the eighteen months that followed levelling the waste land at Criesbach railway station, removing an old vineyard wall, and using the stone as a gravel base for the floor of the hall. Slowly but surely, they erected the 12 x 30 m production facility made of steel and aerated concrete blocks.

During this time, Fritz Müller spent a lot of time travelling around the country on business. To the dismay of his father-in-law, he enticed Gerhard Wippenbeck away from Herion and made him his first sales rep in 1966. Nevertheless, some meetings with customers and suppliers just had to be dealt with by the boss himself. And so it was with the meeting that resulted in GEMÜ and Georg Fischer going their separate ways.

'Business with Fischer had been going very well indeed,' recalls Müller. 'They supplied us with valve bodies and in return we, as agreed, sold them our finished plastic valves. This worked well for about two or three years. Then one day I was invited to a confidential meeting in Schaffhausen where they tried to apply the thumbscrews. They said that in future, I would have to tell them to whom I was selling my valves, which contained Fischer-made valve bodies. They claimed that it all had to do with guarantees, but I immediately realised that they just wanted to get their hands on my customers' contact details. They wanted to muscle in on my business!'

Fritz Müller asked for some time to consider the matter and set off on a long walk to think things over. 'After walking for some time, I sat down on a bench in the vineyards overlooking Schaffhausen,' he recalls. 'I looked down on Georg Fischer's impressive production facilities and decided there and then that I, little Fritz Müller from Ingelfingen, would buy my own injection moulding machine and start making my own valve bodies.'

Back in Ingelfingen, Müller and Helmut Weber set to work designing a suitable mould for the new machine. He then began searching for the best injection moulding machine manufacturers in the business. Once he had identified the top five companies, he made appointments with them for practical tests. He travelled to their premises in Germany and Italy and watched how well the machines coped with making valve bodies using his home-made mould. He eventually decided in favour of a machine constructed by Ankerwerke Gebr. Goller (now Demag Plastics Group) in Nuremberg, which was at the time the world's leading developer of injection moulding machines. At Bürkert, Fritz Müller had learned that it is worthwhile for smaller companies to invest in the very best technology. 'Christian Bürkert always had the most modern plant available. In this regard he was very often one step ahead of his larger competitors,' recalls Müller.

However, this technological advance came with a hefty price-tag: the injection moulding machine cost DM 100,000. The securities required by the bank for a sum this size were correspondingly high. 'I had no choice but to put up everything that we had worked for at GEMÜ. I knew that if something went wrong, we would face a mountain of debts and would have to lay off our staff. That put me under huge pressure,' he says. But Fritz Müller was driven more by his sense of responsibility than by the pressure. 'I did not want to disappoint the people who had placed their trust in me: my family, who had provided me with start-up capital, and my friends, who were working for me although they could have had safer, better paid jobs elsewhere. I wanted to prove to these people that I was worthy of their trust,' he recalls. To ensure success, GEMÜ had to produce more; in order to produce more, it needed new machines, which in turn could only be financed by taking out more loans. 'This meant that the pressure on me was constantly growing and pushing me forward,' says the entrepreneur.

In order to avoid becoming dependent on suppliers, Fritz Müller invested in his own injection moulding machines at an early stage in the company's history

Fritz Müller was passionate about gliding and spent every weekend with his family at the airfield in Niederstetten

In order to cope with the pressure, Fritz Müller needed to relax and switch off on a regular basis. This he was able to do in the air. Since the summer of 1965, he had been spending the weekends with his wife and child at the recently opened glider airfield in Niederstetten. 'Although we worked late on Saturdays, we – and that includes Regina, who had just learned to walk – were at the airfield at 8.00 a.m. on Sunday mornings,' recalls Ingrid Müller.

Soon Ingrid Müller was bitten by the gliding bug and in the summer of 1966, during a brief holiday in Hornberg, passed her gliding test. Back in Ingelfingen, she was keen to demonstrate her ability. One Sunday morning at the airfield in Niederstetten, she was towed into the air in a single-seater glider. Ingrid Müller remembers: 'Back then, there were no female glider pilots at the airfield and consequently no lead cushions or trim weights to make up the weight difference to the men. The aircraft in which I set off had been adapted for a much heavier pilot than myself, but nobody seemed at all bothered about it. It was only when the tow-rope was released and I was alone in the air with a very heavy-tailed glider that was almost impossible to manoeuvre, that everyone suddenly realised that something wasn't right. I knew the whole time that I could crash at any moment.There was no radio on board, so I was entirely alone and had to rely completely on the little gliding experience I had. It was all so utterly terrifying that I swore that if I got down alive, I would never fly alone again. After half an hour I managed – with a lot of luck, I can tell you – to land the glider safely.'

Her husband and other glider pilots who had to watch her desperately trying to control the glider from where they were on the ground, were overjoyed that she had managed to land safely. They encouraged her to take to the skies again as soon as possible in a glider that had been adapted to suit her weight so that she would not develop a fear of flying. But Ingrid Müller was true to her word and from that day on only ever flew in a two-seater glider either with her husband or with one of the other glider pilots. Most of the time, however, she stayed on the ground with the other wives, watching her husband set off and waiting for him to return.

On the Saturday of the 1968 Whit weekend, however, she had to wait a very, very long time. A few weeks previously, Fritz Müller had realised his long-held dream of owning his own glider. Although GEMÜ did not at the time have the 35 employees he had previously thought were necessary to finance such a purchase, although he still faced the pressure of repaying the loans he had taken out for the injection moulding machine, and although the company urgently needed a second car, when he received an offer from Switzerland for an inexpensive second-hand 'Diamant 17,5', he just couldn't resist. The Diamant 17,5 was at the time the best high-performance fibreglass glider on the market by the renowned *Flug- und Fahrzeugwerke Altenrhein* (FFA). Throwing caution to the wind, he bought the glider.

Fritz Müller simply couldn't wait to test his new glider. On the Saturday of the 1968 Whit weekend, he and glider pilot Helmut Laffsa set off from Niederstetten in separate gliders on a comparative flight. The radio system at the Niederstetten glider airfield was able to track their signals as far as Heilbronn, where the signals disappeared. The only thing the team at Niederstetten knew was that Müller and Laffsa had been heading due west towards the Pfälzer Forest. The gliding conditions that day were not particularly good and the forest afforded very few opportunities for landing. When hour after hour passed without a sign of life from either pilot, the ground party in Niederstetten was no longer uneasy, but downright worried.

Meanwhile, Müller and Laffsa were having a ball. Müller remembers: 'The early stages of the flight had been tricky and things were difficult for Helmut Laffsa flying over the Pfälzer Forest in his less-powerful glider. But then we hit amazing thermals and with a stiff easterly wind behind us, we just kept on going. When we flew off our maps, we spoke to each other by radio and agreed to keep course 260° to the West.' Helmut Laffsa landed in the late afternoon at a French military airbase near Reims. Müller continued flying alone and landed at about 6.00 p.m. in a field in the little community of Troyes, more than 420 kilometres from where he had set out.

His first very own glider: Fritz Müller in the cockpit *of his Diamant 17,5*

Laffsa and Müller would have dearly loved to contact their wives immediately after landing and tell the return team where they were, but they were arrested by the police as soon as they landed and interrogated for hours on end. 'It was the end of May 1968, the student revolts had just ended, and there was a general strike across the country,' says Müller. 'The day before we set off, de Gaulle had gone on the radio threatening to impose a state of emergency. So the atmosphere in France was extremely tense. In the middle of all this, two daft Germans arrived in their gliders and landed one shortly after the other without warning not that far from Paris. I, at least, had no identification with me and my glider was still registered in the name of a 78-year-old Swiss woman. It was, of course, all highly suspect. The police probably thought we were some kind of revolutionaries!'

It was 10 o'clock in the evening before the misunderstanding was cleared up and the men were at last allowed to call home. Meanwhile, the party back home was convinced that Laffsa and Müller had crashed. Understandably, when they called to say that they were safe and sound, the relief was immense. The return team immediately set out for France and – after a brief detour to Paris – accompanied the two adventurers and their gliders back to Ingelfingen in triumph.

Fritz Müller and Helmut Laffsa at the end of their long flight to France

CHAPTER 8

CONTINUOUS GROWTH: GEMÜ IN THE 1970s

Although Fritz Müller and Helmut Laffsa had to draw on every one of their skills as glider pilots on the flight to France, their lives had never been in serious danger at any point along the journey. A few months later, however, a seemingly harmless boat trip undertaken by the two men and Laffsa's 12-year-old son almost cost them their lives. Their boat capsized while passing a weir. Laffsa's son managed to swim to the safety of the river bank; Müller also narrowly managed to struggle his way out of the turbulent waters at the weir. Laffsa, however, was unable to free himself from the torrents by himself. When Müller realised that his

friend's life was in danger, he swam back to the weir, managing to pull Laffsa to safety with the last ounce of his strength. Anglers at the scene helped drag the unconscious Laffsa and an exhausted Müller, who had been worn out by his exertions on his friend's behalf, onto dry land.

Life at GEMÜ was much less turbulent. The new injection moulding machine was proving an excellent investment and slowly but surely, GEMÜ plastic valves were replacing other valves wherever industrial applications required aggressive or

SALZBURGER NACHRICHTEN 1968

Paddler in letzter Sekunde gerettet

Mit drei Urlaubern besetztes Paddelboot kenterte auf der Saalach bei Saalfelden + Deutscher in einem Strudel „gefangen" + Bewußtlos von Bekanntem gerettet + 12jähriger rettete sich aus eigener Kraft

In einer dramatischen Rettungsaktion wurde Samstag nachmittag der 41jährige Gipsermeister Helmut Laffer aus Ingelfingen (Bundesrepublik Deutschland) in der Nähe von Saalfelden buchstäblich in letzter Minute vor dem Ertrinkungstod bewahrt. Einem Bekannten Laffers, dem 29jährigen Friedrich Müller, gelang es, den bereits Bewußtlosen aus einem Strudel der Saalach zu befreien und an Land zu bringen. Nach Augenzeugenberichten wäre Müller um ein Haar selbst von den Wassermassen mitgerissen worden.

Helmut Laffer hatte gemeinsam mit seinem zwölfjährigen Sohn Heinrich und Friedrich Müller — die drei verbringen zur Zeit ihren Urlaub im Thumersbach bei Zell am See — Samstag nachmittag eine Paddelbootfahrt auf der Saalach unternommen; der Fluß ist in diesem Bereich ein beliebtes Paddelrevier. In der Nähe der Ortschaft Haid bei Saalfelden geriet das Paddelboot der drei deutschen Urlauber in reißendes Wasser. Das Boot kenterte und die Insassen stürzten ins Wasser. Der zwölfjährige Heinrich Laffer erreichte aus eigener Kraft das Ufer; sein Vater wurde jedoch von den Strudeln buchstäblich gefangengehalten. Müller, selbst schon nahezu am Ende seiner Kräfte, schwamm zu Laffer und zog ihn in ruhigeres Wasser.

Bei der Bergung des bewußtlosen Helmut Laffer halfen auch zwei Fischer, Georg Hutter und Gustav Rodler aus Niedernsill. Die beiden nahmen sofort Wiederbelebungsversuche vor, die nach zehn Minuten Erfolg hatten. Helmut Laffer wurde ins Krankenhaus Zell am See gebracht, konnte aber nach kurzer Behandlung ins häusliche Pflege entlassen werden.

It was on just such a boating trip as the one pictured here that the boat carrying Fritz Müller, his friend Helmut Laffsa, and Laffsa's son capsized at a weir; the three only just managed to avoid being drowned

highly contaminated media to be transported, for example in the chemical industry or in water treatment plant. Fritz Müller remembers: 'When word got around that our valves lasted much longer than our competitors' stainless steel valves and performed just as well, business grew steadily. The best part of it was that because our customers prized durability above all else, they were willing to pay almost the same price for our valves as they were for similar stainless steel models, even though plastic valves are, of course, much cheaper to produce. This ensured us a sizeable margin.'

Nevertheless, Fritz Müller knew that he could not rest on his laurels and had to continue extending his range of products. After all, customers' needs are constantly changing. In the second half of the 1960s, for example, interest in 'diaphragm valves' grew in the industrial sector.

Diaphragm valves are closed using a diaphragm that is pressed against a weir in the valve body, which is positioned accross the direction of flow. The main advantage of the diaphragm valve is that it boasts excellent flow properties. Even if suspended matter or other solid particles accumulate at the weir when the diaphragm is closed, they do not constrict the channel. Moreover, as long as the flow speed is sufficiently high, these particles are swept along by the flow when the diaphragm is lifted once again. This means that diaphragm valves are virtually self-cleaning (as the experts would say, they have 'no deadlegs'), which makes them very attractive for use in sectors that have very high purity and hygiene standards, e.g. the food or pharmaceutical industries. Diaphragm valves are also suitable for use in procedures characterised by heavily contaminated media, e.g. water and waste water treatment or the recycling sector.

These last two areas in particular require corrosion-resistant valve bodies. Until the end of the 1960s, however, only diaphragm valve bodies made of metal or stainless steel – both of which are much less resistant than plastic – were being produced. Fritz Müller saw this gap in the market and knew that GEMÜ could fill it. After the 1968 Hanover Fair,

202 *Funktionsprinzipien und besondere Merkmale*

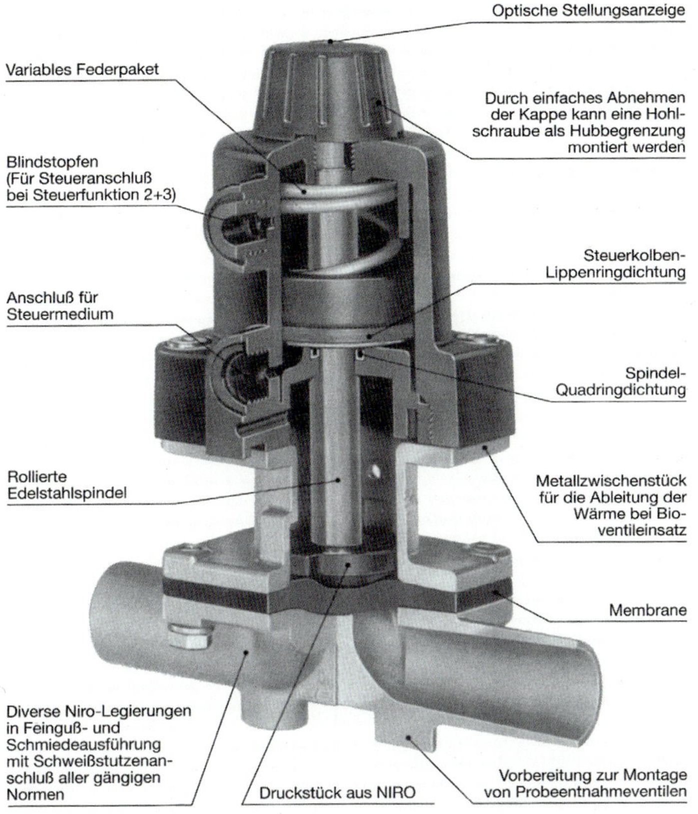

Abb. 3-128. Fremdgesteuertes Membranventil, federkraft-schließend.

Cross-section of a diaphragm valve

Diaphragm valves are not a twentieth century invention: the Romans used simple versions of this valve in the systems that heated their public baths. In the nineteenth century, P. K. Saunders, a South African mining engineer, adopted the operating principle of these simple valves, adapted it for use in the ventilation and drainage systems in African gold mines, and took out a patent on the design. In the decades that followed, a number of mainly North American valve manufacturers obtained licences to produce 'Saunders Diaphragm Valves', which were at the time operated manually. It was, however, only in the early 1950s that industrial companies became interested in this kind of valve: the introduction of pneumatic control systems made diaphragm valves ideal for use in automated production processes. With the market launch of more resistant plastic diaphragms in the 1960s, diaphragm valves went from strength to strength.

which had been a huge success for the company, he and an engineer he had recently hired set about developing the first pneumatically controlled plastic diaphragm valve. A short time later, the company launched the resulting valve (GEMÜ Type 680) on the market. The customers were thrilled. Orders flooded in from all over Europe; GEMÜ's period of growth continued.

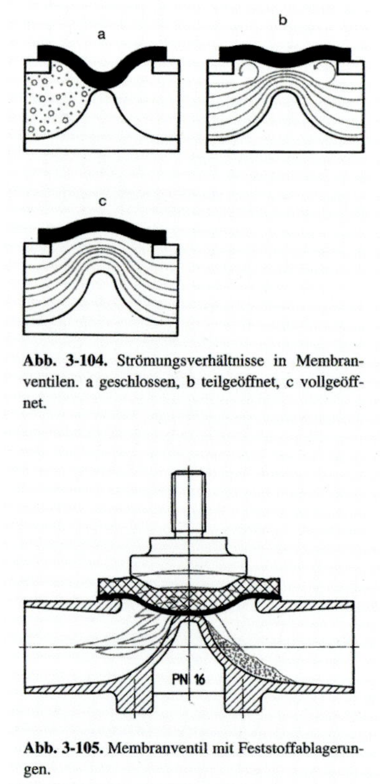

Flow conditions in a diaphragm valve

Fritz Müller's success in business was, however, overshadowed by personal tragedy. Within the space of a few years, Fritz Müller lost both his grandfather and his mother, the two people who had unreservedly supported his plans to set himself up in business. Before passing away at a ripe old age, Friedrich Müller had witnessed the decline of the plastering business he had built up and which had latterly been run by one of his sons-in-law. Marie Müller died in December 1969 at the age of just 65. 'My mother wore herself out on her children's behalf; I am convinced that this is one reason she died so young,' says Fritz Müller. Ingrid Müller agrees: 'My mother-in-law supported my husband and I in everything we ever did, great or small. Without her support, I would never have been able to raise our daughter Regina and work full time for GEMÜ.'

In the early years of the company, Ingrid Müller had shouldered the accounting and secretarial work almost single-handedly. However, the company was growing fast and with it the workload in these areas. The back office was desperately in need of reinforcements. Following the death of her mother-in-law, Ingrid Müller also needed more time to look after her daughter. To lighten her workload in the company, Fritz Müller employed a number of administrative clerks and secretaries.

Here too, Fritz Müller recruited new staff from his circle of friends and acquaintances. 'At that time, I almost always hired people from the locality,' he says. 'If I didn't know the candidates personally, I used a little trick to find out more about them: after the interview, I would take the candidate on a little tour of the company. As often as possible, I stopped

Ingrid and Fritz Müller both worked full time for GEMÜ in the early 1970s; following the death of her mother-in-law and the birth of her son Gert, Ingrid Müller gradually withdrew from the company

at the machines where local staff members were working – people on whose judgement I knew I could rely. Within earshot of these staff members, I would then tell the candidate that I would get in touch in the coming days. About an hour later, I would take another tour of the company, this time on my own. While doing the rounds, I would get chatting to staff members. Then, without being prompted, they would generally tell me what they thought of the candidate. The worst case scenario was that I would hear remarks like: "What? You want to employ him? His father and his grandfather were good for nothing and he is no better, let me tell you!" or "Him? He used to steal chickens as a boy!"

Partial view of the new GEMÜ premises built in 1972

The GEMÜ design department was, like all the rooms in the new building, light and airy

These final pieces of the puzzle helped me to complete the picture. It was a bit sly of me, I admit, but it worked.'

It didn't work all the time, however. In 1971, Müller was obliged to lay off his works manager and replace him with Manfred Hofmann, a former colleague from Herion. Hofmann had left the Stuttgart-based company around about the same time as Müller and had set himself up in business as a freelance precision engineer. 'To be honest, when I worked for Herion, I had my doubts about Fritz Müller and his lofty ambitions. However, in the years that followed, I saw how well GEMÜ had developed. So when Fritz Müller came and offered me the job of works manager, I was confident that GEMÜ was strong enough to withstand periods of economic instability,' recalls Hofmann. 'I had a wife and children at the time and was not as willing to take risks like I was when I first went freelance; the prospect of having more security than I had on my own was a huge draw. That being said, I would not have given up my freelance work had I not felt sure that Fritz Müller would give me the freedom I needed to work and act independently as his works manager. In short, the job he offered me was the ideal combination of independence and security.' Hofmann brought with him his machines, tools, and above all his technical expertise and industrial experience. In the years that followed, he was instrumental in adapting the largely manual production processes he met with on his arrival at GEMÜ to meet the needs of serial production.

As the company continued to expand, the production unit once again grew too big for the hall in which it was housed. 'When 15 of us moved into the production facility at Criesbach railway station in 1965, I said to Helmut Weber, "We will know that we have made it in this business when we outgrow that hall."' recalls Müller. 'In 1968, we had to build an extension, and less than four years later, GEMÜ had been growing so fast that, once again, we didn't know where to put all the people that worked there.' The administrative section of the company – which comprised Fritz Müller's office, the design department, the secretariat, the commercial

Neuer Wasen

department, and the purchasing department – was squeezed into an area measuring 50 square metres. Things were just as tight in the production facility. In order to ensure that production continued smoothly, an extension was urgently required.

In 1972, new premises with a larger production facility and a three-storey administration building were erected on the site in Criesbach. Fritz Müller's office was situated on the top floor directly alongside the 200-square metre flat that now became home to his wife, daughter, and young son Gert. The two storeys beneath this flat housed spacious, bright, and modern offices – the legacy of Ingrid Müller's early days at the company, when she yearned for sunshine and fresh air during the long days spent toiling away in her stuffy little office in the house on Hundsbergweg. A computer simplified and accelerated the handling of orders and made it easier for the recently established marketing department to assess and analyse figures. Employees had a canteen in which they could eat their meals and a sauna in the basement, where they could relax and unwind at the end of the working day. The team moved into the new building in the summer of 1973, just in time for the tenth anniversary of the company's establishment, which the Müllers celebrated with their over 100-strong workforce.

By now, GEMÜ was producing not only large quantities of plastic valves, but also a wide range of valve component modules that could be combined to meet the customers' individual needs. Customised valves were also made on request. The development department worked hard to realise ideas and suggestions relayed by customers to the company via GEMÜ's sales representatives. Right from the word go, Fritz Müller had decided not to hire straightforward sales reps, opting instead to fill these positions with qualified designers or engineers. 'I wanted my customers to have reps with whom they could talk and who understood what they were talking about,' says Müller. At the same time, the qualified sales engineers were able to explain in detail to the development department what the customer really wanted. This direct flow of information

10 years of progress and growth: Fritz Müller speaks at the anniversary of the company's foundation

made it possible for GEMÜ to launch new products on the market quickly. Once a new product was ready for launch, the product managers began searching for other areas of application that would allow the new model to be produced in larger, less expensive batches.

Just how quickly and effectively the sales and development teams worked together was illustrated by GEMÜ's development of metal valves. In the late 1960s, a customer called for valves that could be used to transport vaporous media. In view of the fact that GEMÜ's plastic valves could only be used to temperatures of up to 80 °C, GEMÜ developed a valve with a metal valve body for this customer. In the years that followed, GEMÜ continually extended its range of metal valves at the instigation of the product management team. While searching for new applications for the metal valves the company was producing, GEMÜ's product managers lighted on polystyrene production in the early 1970s. Although polystyrene had been invented in the early 1950s, it really only took off on the market in

Even in the early 1970s, GEMÜ employed qualified engineers as sales representatives

the late 1960s. In some production procedures for this material, a hollow mould is filled with small, hard balls of polystyrene and subjected to heat in the form of hot vapour. The application of heat makes the polystyrene balls expand and bond with each other. Heat-resistant metal valves for vaporous media were needed for this expansion process. However, the globe valves that GEMÜ was producing in the early 1970s did not have a sufficiently high throughput for this purpose; globe valves with an angled seat were needed. In order to meet this demand, GEMÜ developed the first metal angle seat globe valves in 1972.

Until the end of 1972, most of these valves with their hot-pressed brass, stainless steel, and diecast aluminium metal components were made by GEMÜ's partner, Herion, which held a stake in the company. Relations between the two companies had never been easy. Erich Herion's promises that GEMÜ could have its old machines and that the two companies would co-operate on material procurement were never kept. Moreover, sales of GEMÜ valves through Herion were rather sluggish. Nevertheless, Herion was keeping a close eye on GEMÜ's economic development. Müller remembers: 'Once a month, a controller came from Stuttgart and asked me to show him every one of our receipts. Me! Who isn't the least bit interested in accounting and just wanted to make valves! I hated being monitored in that way.'

What was even more problematic was the fact that Fritz Müller had to finance all investments in the company out of his share of the company profits, while Erich Herion left all his profits in the company, thereby continually increasing his stake. 'I felt like I was digging my own grave,' recalls Müller. 'The harder and faster I worked, the more products I launched on the market and the more profit I made, the smaller my share in my own company became. I knew that I would some day reach the point where Herion would close the trap around me and say: "So. This is now my company. From now on, you are working for me!"'

Although this was an unbearable prospect for Fritz Müller, he could not see a viable way out of the contract. Eventually matters took their own course. Following the death of Erich Herion in the spring of 1968, his share in GEMÜ passed into the personal ownership of his son, Erich Herion Jr. However, because the articles of association did not permit associates to hold such private shares in the company, when Herion's heirs began to fight about the inheritance, Erich Herion Jr was forced to sell his stake in GEMÜ. Fritz Müller grabbed this unique opportunity to buy him out. 'I scrabbled together all the money I had and took out a loan. I was forced again to put up everything we had as a security for the loan. I remember it well: shortly before Christmas 1971, we processed the required sum of money for payment. My accountant made the payment, turned around to me, and said, "Mr Müller, that's it. We now have nothing left."' Fritz Müller begged to differ. He was, at long last, head of his own company again. Based on his experience with Herion, Müller decided never again to have associates or partners in his company. 'At the time, I used to say to myself: "better to have a loaf all to myself than have to share a cake" and I have abided by that principle to this day,' he says.

In the years after Fritz Müller bought back Herion's stake in the company, GEMÜ proved Manfred Hofmann right: it was indeed a company capable of withstanding an economic crisis. While Germany went into recession in the wake of the 1973 oil crisis, GEMÜ continued to go from strength to strength. As a supplier of valves for water treatment plants, it benefited from the growing awareness about environmental protection and pollution triggered by the green movement and Germany's tough environmental regulations. At the same time, GEMÜ succeeded in winning new customers in the chemical and food industries. Finally, the spread of microelectronics and the associated automation of industrial production processes created a steady stream of new application opportunities for both plastic and metal valves.

In order to keep pace with these developments, GEMÜ had to invest in the very latest production technology. The first step was to buy the machines and tools needed to produce metal valves entirely at GEMÜ. These purchases were followed by the procurement of machinery for the manufacture of new GEMÜ products. Fritz Müller also invested simultaneously in development. In 1977, GEMÜ diversified into measurement systems with the launch of models 800 and 850, two plastic flowmeters with floats and graduated tubes made of polysulphone and Trogamid, which were more durable than the glass tubes that had been used up until that point. In 1978, the Ingelfingen-based company launched no less than three new globe valves (models 512, 514, and 554) on the market. They were fitted with pneumatic actuators made of plastic or aluminium, which made them lighter and less expensive than the conventional metal actuators in use at the time.

Lord of the manor: Fritz Müller in his office (c.1973)

In order to ensure that the customers were properly informed of the benefits of these products, GEMÜ's sales engineers received continual training. But they were not the only ones to maintain close contact with the customers, Fritz Müller was determined not to lose touch with the market. He strengthened business relationships by going to see the customers in person, testing new machines, and attending the major industrial trade fairs in order to ferret out new applications for GEMÜ valves. 'Even at that stage I could have said to myself, "Why spend your days and nights working, you have your designers, sales engineers, commercial staff to do it for you?"' says Müller. 'You can do that. But if you do, things will only run smoothly for a time. Soon there will be a lack of impetus. Those who do not continually improve, find themselves edged out of the market.'

Fritz Müller completed many of his business trips in his own plane. Although his heart remained true to gliding, having reached a height of 8,000 metres while flying over Lake Constance in 1970 – thereby earning himself the diamond version of the 'golden badge of achievement', the highest accolade in German gliding – he was looking for a new challenge. In 1974 he got his pilot's licence and a year later, he bought himself a second-hand single-engine Cessna for DM 60,000. 'I was increasingly away on business trips,' recalls Müller, 'and it took ages to drive to the airport in Stuttgart or Frankfurt. If only to save time, it was worth flying to meetings in my own plane from Schwäbisch-Hall.' But Müller used the Cessna for more than just business; he often took

Fritz Müller with passengers on board his Cessna (mid 1970s)

Fritz Müller and his son Gert on a weekend excursion to Lake Constance (late 1970s)

the family out on little excursions to escape the humdrum of daily life. 'Sometimes we would spontaneously set off on a Sunday afternoon and fly somewhere, mostly to Sylt or to Lake Constance. We would spend the afternoon there, go out for a nice meal, and return home late in the evening,' recalls Ingrid Müller.

CHAPTER 9

'NEW PRODUCTS FOR NEW MARKETS': GEMÜ'S INTERNATIONAL EXPANSION (1981–90)

The 1980s got off to a promising start: GEMÜ was unaffected by the recession in the German economy after the second oil price shock of 1979/80. The Ingelfingen-based company exhibited successfully at the IFAT in Munich, the world's most important trade fair for the waste disposal sector that takes place every three years. It also won renowned prizes for outstanding industrial design at the Hanover Fair in 1981. With a workforce of just under 200, GEMÜ was now producing over 60,000 valves per annum, of which approximately 30 per cent were destined for export. Most of these exports were sold by agencies in Western Europe.

Although the company was already exporting its products, Fritz Müller could see that there was room for improvement in its international sales structure. Prior to 1971, GEMÜ's exports had been taken care of by the Swiss company Bachofen, which represented a large number of other companies in addition to GEMÜ. Furthermore, unlike GEMÜ's sales representatives in Germany, Bachofen's reps were not trained sales engineers. If they were able to increase their margins by selling products that required less effort and explanation than GEMÜ valves, they preferred to do so. Says Müller: 'GEMÜ was simply not Bachofen's priority. We were just one of many. I, on the other hand, wanted GEMÜ to be Number One and I wanted our valves to sell.' When the opportunity arose in 1971 to buy Bachofen's sales companies in France, England, Spain, and Scandinavia, Fritz

Design awards for GEMÜ products (1987): to this day, GEMÜ regularly takes part in design competitions and wins awards

Auszeichnung

2/2-Wege-Kunststoff-Klappe,
fremdgesteuert
- GEMÜ 410 -

Dieses Produkt
wurde von einer
internationalen Jury
für die Ausstellung

if

Die gute Industrieform

auf der
Hannover-Messe
ausgezeichnet.

Hannover, im März 1987
Der Vorstand

Sepp D. Heckmann

Ernst Josef Auer

Auszeichnung

2/2-Wege-Kunststoff-Membranventil,
fremdgesteuert
- GEMÜ 610 -

Dieses Produkt
wurde von einer
internationalen Jury
für die Ausstellung

if

Die gute Industrieform

auf der
Hannover-Messe
ausgezeichnet.

Hannover, im März 1987
Der Vorstand

Sepp D. Heckmann

Ernst Josef Auer

Auszeichnung

2/2-Wege-Metall-Membranventil,
motorgesteuert
- GEMÜ 698 -

Dieses Produkt
wurde von einer
internationalen Jury
für die Ausstellung

if

Die gute Industrieform

auf der
Hannover-Messe
ausgezeichnet.

Hannover, im März 1987
Der Vorstand

Sepp D. Heckmann

Ernst Josef Auer

Auszeichnung

Durchflußmesser nach dem
Schwebekörperprinzip
- GEMÜ 850 -

Dieses Produkt
wurde von einer
internationalen Jury
für die Ausstellung

if

Die gute Industrieform

auf der
Hannover-Messe
ausgezeichnet.

Hannover, im März 1987
Der Vorstand

Sepp D. Heckmann

Ernst Josef Auer

Müller did not hesitate. He bought the companies and integrated them into a specially founded GEMÜ subsidiary, InterApp GmbH, which had its headquarters in Holzhäusern, near Zug in Switzerland.

Although InterApp did have a small production unit for tax reasons, the main purpose of the company was to manage GEMÜ's foreign business. However, after the foundation of the company, it quickly became apparent that InterApp did not have the necessary expertise. That being the case, Fritz Müller was delighted to learn that Hanspeter Stoll, the man who had been his main contact at Bachofen, was interested in coming on board. Soon after the establishment of InterApp, Hanspeter Stoll switched from Bachofen to GEMÜ's new sales company and from that day on, managed GEMÜ's entire foreign business in Europe. The co-operation between GEMÜ and InterApp progressed very well indeed. Nevertheless, as the years passed, both companies began to yearn for greater independence. In 1981, the two companies separated. Hanspeter Stoll paid 1.5 million Swiss Francs for InterApp and the

The premises of GEMÜ GmbH in Rotkreuz, the Swiss subsidiary established in 1981

right to continue selling GEMÜ products worldwide except in France, Scandinavia, Great Britain, Austria, and Brazil. Fritz Müller used the proceeds of this sale to build up a production facility at GEMÜ's recently founded Swiss subsidiary, GEMÜ GmbH, which, unlike InterApp, operated solely as a production company. From this point on, GEMÜ's foreign business was taken care of by subsidiaries that Fritz Müller established almost simultaneously in France, Great Britain, and Sweden.

This new sales strategy made it possible for Fritz Müller to forge closer relationships with his customers. 'If a company is not actually present on a market, it runs the risk of losing the direct contact with that market,' explains Müller. 'But I want to be as close to the customer abroad as I am in Germany. I want a direct line to the market so that I know what products are in demand and what prices are being paid on that market, regardless of the fact that customers generally seek to make direct contact with the manufacturer. This is why, since coming to this realisation, I have always founded subsidiaries wherever it is worthwhile.'

Such a strategy requires patience and staying power. 'In the early years after the establishment of a foreign subsidiary, the success of that subsidiary is generally relatively modest,' explains Müller. 'One has to take that into account, because the market only ever opens to us to the degree that our organisation develops. In the early years, it is all about investment and building up a company; earning money comes later.'

This principle proved well-founded when it came to the most important GEMÜ project of 1981, namely the establishment of GEMÜ's Brazilian subsidiary. While GEMÜ products were already well established on the markets of Europe, the American continent was largely uncharted territory for the brand, and that was exactly what attracted Fritz Müller. 'Initially, I was just thinking in terms of "overseas trade". I hadn't yet decided where exactly "overseas" was or whether I wanted to focus on North or South America,' he recalls.

In order to get a picture for himself, Müller travelled across the Atlantic and spoke to representatives of a number of banks and chambers of industry and commerce in various large cities in Canada and the USA. 'Of course,' he says, 'each of them wanted to paint the most flattering picture possible of their city. If you fly to Vancouver after a meeting in Montreal, they'll tell you: "Forget Montreal. It's much too cold there; it's much better here." If you fly to New York, they'll tell you: "You know, we are the largest city on the eastern seaboard. This is where the market is." If you fly to the west coast, to San Francisco, they'll tell you: "New York is riddled with trade unions. Save yourself the bother; come to the west coast." And so on and so forth.'

Fritz Müller travelled on to Brazil, continuing his search for the ideal location for his American subsidiary. Brazil's economy had been posting growth rates of up to 10 per cent since the 1960s. 'We started out in the north, in Bahia. There was nothing there but caipirinha, beaches, football, and women. Not for us. Rio was exactly the same, but with a bit more class.' Sao Paulo, the next stop on Fritz Müller's reconnaissance tour, was the hub of the Brazilian market, but the 20-million strong metropolis with its vast slums also had the highest crime rate in the country. Although the affluent and beautifully situated Porto Alegre in the south didn't have these problems, it was situated 850 km away from Sao Paulo and was, for this reason, practically cut off from the market. Disappointed, Fritz Müller flew back up north. Purely by coincidence, he found the right location for his overseas subsidiary during a stopover in Curitiba. 'I knew when we were coming in to land that this was it!' he recalls.

Initial meetings with economic representatives of the city backed up this good impression. Curitiba was so well situated in terms of transport connections, that Sao Paulo – although at a distance of 450 km – was only a few hours' drive away. With less than two million inhabitants, the city was exactly the right size; it also had a good administrative system and low crime rates. Another point in the city's favour was the fact that a number of German companies had already set up branches there. Fritz Müller decided to base his overseas business in Curitiba.

Above left:
Initially, GEMÜ Brazil was just one of many companies in the industrial estate in Curitiba

Above right:
Fritz Müller (left) and Herbert Neubauer (right), the man responsible for building up GEMÜ Brazil, in the company canteen

In view of the fact that the Brazilian authorities impose high import duties, there was no question of simply setting up just a sales company. The plan was that GEMÜ Brazil would manufacture products for the Brazilian market. Müller decided to start by producing plastic valves that would carve out a niche for the brand on the Brazilian market. Initially, everything went according to plan. Criesbach railway station had not been in use for many years and some of the tracks and sidings were on GEMÜ's property. Fritz Müller had a railway wagon brought onto the siding so that it could be loaded with machines, tools, and assembly equipment. The Kocher Valley railway line – which at the time still ran on a regular basis between Waldenburg and Forchtenberg – passed GEMÜ's premises, and the intention was to hook the loaded wagon onto one of the trains. 'The plan was as follows: we would transport the machinery to Hamburg by train, ship them to Brazil, unload them in Curitiba, and build up production from there,' says Fritz

Müller. On the day before the wagon was due to set off, a fax arrived from Brazil: before issuing an import permit, the Brazilian customs required a written assurance that GEMÜ Brazil would post turnover of at least 12 million dollars in the first business year. Otherwise a hefty fine would be imposed. For Müller, this was out of the question. 'I called an immediate stop to the transportation. The machines were unloaded, brought back to the production facility, and used to manufacture products for the German market from then on.' This was the end of the plan to conquer the Brazilian market with GEMÜ plastic valves.

It may have been a setback, but it was not the end of Fritz Müller's Brazilian dream. He changed his strategy and decided to launch production in Brazil with the manufacture of simple, all-purpose valves made of cast iron that local foundries could produce using GEMÜ designs. There was always a demand for such valves, he reasoned; they would allow GEMÜ to gain an initial foothold on the Brazilian market. Müller had not planned any further than that. He instead set to work redesigning successful GEMÜ valve types so that they could be made of cast iron. Müller entrusted Herbert Neubauer with these cast models and DM 300,000 start-up capital. The two men knew each other from flying and Müller had hired Neubauer to build up and manage the recently established GEMÜ Indústria de Produtos Plásticos e Metalúrgicos Ltda. in Curitiba.

The start-up period was, however, more difficult than anticipated. 'I approached the whole thing in a very naïve way,' says Müller. 'The 1981 Hanover Fair was under the patronage of Brazil. There was much talk throughout the fair about the huge opportunities that awaited foreign investors in Brazil and its huge, supposedly undeveloped industrial products market. I had been in Brazil myself only a short time previously and should really have known better, but somehow I had this image in my head that the country was full of Indians and gauchos. Of course that was not the case at all. There was not only a lot of industry in Brazil, but also a few major valve manufacturers who were not entirely thrilled to see GEMÜ muscling its way on to their traditional hunting ground. The

Managing Director of one of these local manufacturers told me to my face that he would make sure we didn't survive the first year in Brazil.'

Initially it seemed as if the competitor in question had nothing to worry about. 'The first hall we rented in Curitiba was so dilapidated that it quite simply collapsed a year later during a heavy thunder storm. What's more, the DM 300,000 start-up capital was swallowed up in only six months. It had not even been enough to buy equipment for the production facility,' relates Müller. But he refused to give up. 'We had achieved something: we had made contact with some customers, we had a good foundry, and we had hired a few people. I knew that if we stuck with it at that point, things would work out in Brazil. So I injected more capital.'

This high-risk investment was made possible by GEMÜ's solid earnings in Europe. Then, in mid-December 1981, something happened that made it look like this period of stability was over for good. Ingrid Müller remembers: 'During the night of the 15/16 December 1981, we were woken from our sleep by a loud explosion in the production facility. My husband rushed to the window and said: "There is a strange flickering light coming from the hall; I think it's on fire." He ran downstairs to find out. In the meantime, I woke the children and packed our bank documents and any absolute necessities as fast as I could. Then everything happened very quickly: the material for the plastic valves had burned like tinder and by the time my husband had come back to get us out, the stairwell had already filled with toxic smoke.'

The Müllers just managed to get out in time. Ingrid Müller and the children were brought to hospital with suspected smoke poisoning and spent the night with the family of Karl Dörr, a long-serving GEMÜ employee. Despite the fact that it was minus 16° outside, Fritz Müller stayed at the company and spent the night watching the firemen quench the flames. When the sun came up the next morning, he was faced with the ruins of his company. Smoke and the water from the firemen's hoses had caused serious damage to the administrative building and the Müller's private flat; the production facility was almost completely burnt out; most of the machinery was only fit to be dumped. The damage caused by the fire, which had been started by the carelessness of two burglars, amounted to almost DM 15 million. Even though he knew that the cost of rebuilding the company would be borne by the insurance company, Fritz Müller had no idea how he was supposed to keep production going. 'In the days following the fire, I saw my husband completely despondent for the first and only time,' says Ingrid Müller.

Things could, however, have been a great deal worse. In 1978, immediately after joining the company, the new commercial manager, Eckhard Noller, resisted Fritz Müller's opposition and took out business interruption insurance. This expensive optional insurance now turned out to be the company's saviour. The insurance company's experts sat down with Fritz Müller and

'Completely despondent for the first and only time': Fritz Müller and employees on the grounds of the company after the fire that destroyed much of the premises

his staff and drew up a concept that would get GEMÜ back on its feet and in production again as quickly as possible. It also made the necessary funds available. While the rebuilding of the premises got underway in earnest, a 200-square metre inflatable tent was used as a temporary production facility on the banks of the river Kocher.

On 16 December 1982, exactly one year after the fire, GEMÜ moved into its new premises. Some 1,000 square metres larger than the previous complex, the new premises had been built with growth in mind. At the ceremony to mark the opening of the new premises, Fritz Müller thanked above all the employees that had remained loyal to GEMÜ during the rebuilding period and had kept production up and running even under the most difficult circumstances. Thanks to their tireless efforts, the Ingelfingen-based company managed to increase

The GEMÜ production facility on the day after the fire

While the clean-up continued, production continued in a provisional production facility

turnover by 10 per cent in 1982 despite the damage caused by the fire.

GEMÜ posted similar growth rates in the years that followed, despite the fact that market conditions were tougher as a result of competition from American and Japanese companies and rising raw material prices. Fritz Müller rose to these new market challenges by launching new products. In doing so he focussed in particular on measurement, control, and regulation systems. In 1982, GEMÜ launched a pilot valve for pneumatic actuators on the market. A space-saving battery mount version of this brand new valve type was soon launched and others were added, creating an entire series (Types 320, 322, 324, 326). In 1984, GEMÜ presented its first motorised actuators for diaphragm valves. In the following year, the companies' numerous activities in the fields of measurement, control, and regulation systems were combined into a single department.

Fritz Müller's strategy was working. GEMÜ was not only successful on the market, it was also growing. More staff was needed to cope with the constantly rising number of incoming orders. However, because of the increasing number of CNC machine tools (computerized numerical control), which had been introduced in the early 1980s, the search for new staff was much harder than it had been in the past, the reason being that the operation of these tools required different qualifications. These machines, which were capable of manufacturing highly complex workpieces automatically could not be operated by the kind of semi-skilled labourers that had been working in the production unit at GEMÜ well into the 1970s. GEMÜ now needed highly qualified skilled workers with a knowledge of CNC, and such workers were hard to come by. This was certainly the case in the Hohenlohe region, where wages were approximately ten per cent lower than the national average. This meant that the few skilled workers that did live in the region generally went elsewhere to work.

This state of affairs changed GEMÜ's employment policy dramatically. Fritz Müller could now no longer hire his staff from the neighbourhood. Gone were the days when everyone knew everyone else at GEMÜ. Fritz Müller now had to look for employees all over Germany. The drawback was that the Hohenlohe region and the Kocher Valley were relatively unknown at national level and were considered quite unattractive. The situation was further compounded by the fact that an increasing number of young people preferred to pursue a career on the commercial side of business. 'People always think office jobs are better,' says Müller with regret. 'It's such a pity; working in production is certainly not as dirty or physically strenuous as it used to be. Nowadays we need people who can use their heads in production. You have to be on your toes and react quickly. That's what counts.' Despite these difficulties, GEMÜ hired about 15 new members of staff every year in the latter half of the 1980s. When the production unit once again grew too big for its hall in 1986, Müller decided to invest in a major new company building rather than just extending his fleet of machines. He invested DM 5 million in the construction of a further 2,700 square metres of production space in Criesbach. The heart of the new building was a 15-metre high high-bay warehouse in which valve and plant components were stored in groups. This new addition accelerated the provision of parts to the production unit considerably.

Although the intention was that the new building would be ready in time for GEMÜ's 25th anniversary in 1988, there were delays in the construction work. As soon as it became clear that the complex would only be ready in 1989, Fritz Müller promptly decided to postpone the company jubilee by a year. 'Although that wasn't really correct historically,' says Fritz Müller with a grin, 'it meant that I would be able to celebrate my 50th birthday at the same time.'

The modern high-bay warehouse was the heart of the new building

As it turned out, 1989 was certainly a year of celebrations and honours for Fritz Müller. Things got off to a start on 2 January 1989 with a party and reception at Castle Ingelfingen to mark his 50th birthday. In September of the same year, he was presented with the Wolf-Hirth Gold Medal, the highest honorary accolade presented by the Air Sports Association of the state of Baden-Württemberg. For his part, Fritz Müller presented the Kocher Valley Air Sports Club, of which he became president in 1977, with a high-performance LS4 glider, which the members of the club promptly named 'Fritz Müller'.

Having moved into the new company building, GEMÜ celebrated its 25th anniversary with a party for friends and business partners in the Criesbacher Kelter on 14 September 1989 and a party for all employees on the following evening. One of the highlights of the show, which was organised and performed by members of staff, was the song 'Let Müller get on with it', which was composed, sung, and even recorded as a single by employees. To the rhythm of a foxtrot, The GEMÜ Singers, as the group called themselves, reviewed in this song the most important events in the company's history, praised Müller's leadership qualities, and teased him about his beloved 'duck' (a Gyroflug-SC01-Speed-Canard plane he had bought in 1986). The 'duck', which got its nickname from the plane's unusual shape, gave the appearance of flying backwards and regularly caused a stir and was a source of great amusement at air shows.

A huge attraction at air shows: Fritz Müller's beloved 'duck'

The GEMÜ Show Troupe, which played music from every country in which the company was represented, also went down a storm. They certainly had a lot of different music to play that evening. After all, in the course of only eight years, GEMÜ had developed into a successful international company that generated half of its turnover abroad. During this period of expansion, the main focus had still been on the European market. However, once GEMÜ had overcome its teething problems in Curitiba, the Brazilian subsidiary quickly began conquering the South American market. By 1989, it employed approximately 100 of the more than 300 people who worked for GEMÜ around the world.

Encouraged by the company's success in Brazil and dissatisfied with the performance of its representative in the USA, Fritz Müller decided it was time for GEMÜ to break onto the American market in 1988. He chose Atlanta as the location for GEMÜ Valves Inc. This metropolis in the 'Deep South' was one of the last key hubs on the American market. Another advantage was the fact that German companies could rent inexpensive offices with switchboards with the help of the German-American Chamber of Commerce. Initially, the intention was that the US subsidiary would be no more than a sales company. However, it proved very difficult to sell German valves in the USA because of the different standard dimensions

Rapid growth meant that GEMÜ's Brazilian subsidiary had to move into a new building in the late 1980s

Made in Brazil: the subsidiary in Curitiba also supplies GEMÜ's sales company in the USA with valves

in use there. Furthermore, American customers attached less importance to the appearance of valves than the Europeans did and were willing, for example, to do without the external polishing of the valves, which was not essential for the technical function of the product. This is why GEMÜ Valves Inc. began to build up its own production facility for the US market, which developed slowly but surely in the years that followed.

But GEMÜ was not just expanding on the American continent. In the jubilee year of 1989, for example, Fritz Müller founded a subsidiary in Australia. Prior to this move, GEMÜ's Australian customers had been dealt with by the British subsidiary. 'The distance between Europe and Australia was simply too big for the British office to be able to do justice to the Australian market. I had learned from our success in Brazil that it really was worth producing locally and being present on the market. Moreover, the subsidiary in Australia brought us one step closer not only to the New Zealand market, but to Oceania as a whole,' explains Müller.

The figures show how right Fritz Müller was to opt for international expansion: in 1989 GEMÜ sold 290,000 valves and 46,000 instruments for measurement and regulation applications. In December 1989, the company's new commercial director, Michael Nagl, announced that he intended to double the amount of valves and triple the amount of measurement instruments sold per annum by 1994. Given the company's rapid growth in the previous years, it did not seem too ambitious a target.

GEMÜ

A FEW WORDS ON ...

THE PROMOTION OF VOCATIONAL TRAINING

GEMÜ has always attached great importance to the promotion of vocational training. 'I myself enjoyed an excellent education and I would like to give others the chance to benefit from this privilege,' says Fritz Müller. As soon as operation schedules at GEMÜ permitted it, he began taking on apprentices at the company. Initially, only two apprenticeship places were made available every year and the successful applicants received their training in a corner of the production facility. This was not good enough for Fritz Müller. When constructing a new company building in the late 1980s, he invested DM 1.2 million in a modern training workshop. To this day, the equipment in the workshop is regularly updated to ensure that it is state of the art. The training workshop now encompasses a number of laboratories where apprentices learn about pneumatics, hydraulics, and electronic controls, as well as a classroom equipped with the most up-to-date communication technology. In 1989, for the first time, ten young people began apprenticeships at GEMÜ under the guidance of instructors Wolfgang Wick (construction of metal machines and special purpose machines) and Alois Walter (electrical engineering). Since then, the number of apprentices taken on every year has been increased to 16.

Alois Walter and Wolfgang Wick talking to an apprentice in GEMÜ's modern training workshop

In the course of their three-year standard apprenticeship, they receive training in CNC, CAD (computer-aided design), and CAM (computer-aided manufacturing) machines. GEMÜ currently trains young people as tool mechanics, electricians

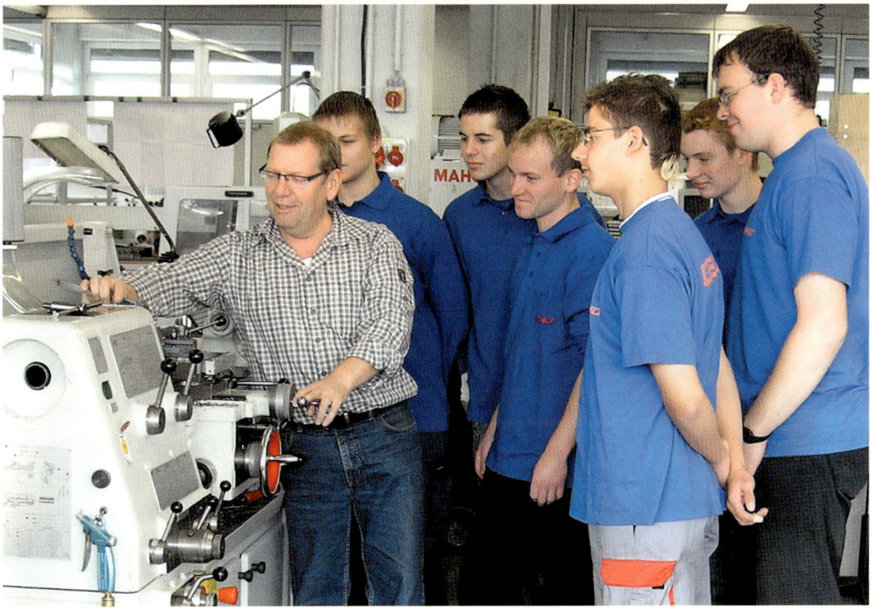

for equipment and systems, process mechanics for plastics and rubber, and machine and plant operators. On the commercial side of things, it also trains draughtspersons, skilled workers for warehouse logistics, and industrial clerks.

Almost all of these apprenticeships are expressly also open to young people who have attended a *Hauptschule* (a German secondary school that provides basic education of neither a vocational nor academic nature up to the age of about 15). 'I myself went no further than getting my elementary school leaving certificate at the age of 14,' relates Müller, 'so I know from experience that the kind of school leaving certificate you have says little about your interests or talents.' He is dismayed at the declining interest among young people in pursuing a technical career. 'We don't just need commercial staff and office staff, we also need intelligent people who have learned a trade,' he says with conviction. This is why, in the late 1980s, he and the headmaster

Former GEMÜ employees and apprentices get involved in one of the company's numerous school projects

of Ingelfingen's combined primary and secondary school, the state school authority in Bad Mergentheim, and the school authority for higher secondary education in Stuttgart launched what was known as the 'Kocher Valley Pilot Project'. This project gave interested young people the chance to complete an 18-day programme including several days of practical work, thereby giving them an idea of what it was like to work as a toolmaker or mouldmaker. Just under ten years later, GEMÜ provided students at the Technische Gymnasium in Crailsheim (a secondary school offering a high level of technical and academic education) with components and advice for a school project. 'We are involved in this project because we consider it very important to create the right basis for technical professions at an early stage,' said Müller at the time.

Despite these isolated efforts, nothing has as yet succeeded in stopping the decline of interest in technical professions. In the past five years in particular, the Heilbronn-Franken Chamber of Industry and Commerce has noted an increasingly marked lack of skilled workers in the Hohenlohe district. This situation is expected to deteriorate further when those young people who were born more recently in years of low birth rates enter the job market. This is why initiatives are being launched all over the region in the hope of fostering the interest of children and young people in technology at the earliest possible age and telling them about opportunities for training in the technical sector.

The fascination of technology: young people at the GEMÜ stand at a careers fair

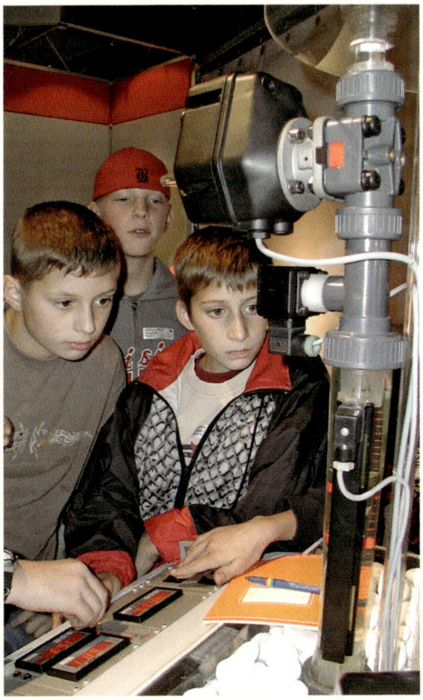

GEMÜ is actively involved in these initiatives. As a member of the Kocher Jagst Innovation Region initiative, GEMÜ has been supporting technical school projects at Ingelfingen's combined primary and secondary school since 2007, making its apprentices available as 'youth tutors'. One feature of these projects is that the pupils do not simply absorb the material passively, but actively acquire knowledge by actually completing experiments themselves. For their part, the work with school children has taught GEMÜ's apprentices how to pass on the knowledge they have gained. GEMÜ is also involved in a project known as the 'Pupil-Engineer Academy' (SIA), which seeks to promote technology classes in the final years of tuition at Gymnasien (academically-oriented secondary schools). For this project, groups of 16-17-year-old pupils from a number of schools work on a complex technical project over the course of a full school year. To complete their projects, they are given access to the production facilities and the expertise of the companies involved in the programme. In addition to these initiatives, GEMÜ regularly offers work placements to pupils, supports events that seek to provide schools in the region with information about technical professions, hosts such events on its own premises, and exhibits at regional careers fairs.

Participants in the GEMÜ training workshop on Girls' Day

GEMÜ also gets involved in 'Girls' Day', an annual series of events that takes place nationwide and seeks to foster girls' interest in entering technical professions. On such a day, 10-16-year-old girls can spend the whole day in GEMÜ's training workshop trying their hand at technical work and creating a workpiece under expert supervision. Fritz Müller is a vigorous advocate of encouraging more women to enter the technical professions. 'At GEMÜ, women of all ages operate machine tools and work in the tool store and tool manufacture unit. With today's CNC technology, it is mental agility, not strength, that counts when operating machinery. The same is true in the development units. My new flow technician is so adept at making computer calculations that she is way ahead of me and some of the other long-standing members of staff,' he says with admiration.

Young men and women who opt for an apprenticeship at GEMÜ are welcomed to the company with a special week of introductory activities. They then spend the first year of their apprenticeship learning the basics of their future profession. At the end of this year, they continue their training at a number of stations in the company, thereby familiarising themselves with the various stages of work in their specific area of expertise. They can also quite literally broaden their horizons by completing work placements at GEMÜ's foreign subsidiaries.

Apprentices at GEMÜ regularly take part in the German national 'Young People Research' competition. In 2007, for example, GEMÜ apprentices won first prize in this competition for developing a remote-controlled brake system for a child's

GEMÜ apprentices win first prize at the German national 'Young People Research' competition in 2007

bobbycar. GEMÜ apprentices also regularly come top of the class in final examinations organised by the Industry of Chamber and Commerce. GEMÜ also rewards the performance of its apprentices: since April 2004, the company has nominated an 'apprentice of the month', the winner of which gets to use a small company car free of charge for four weeks.

Fritz Müller attaches great importance to the fact that apprentices show independence and initiative. 'Not everyone is willing or able to set themselves up in business,' he says, 'but companies also need people who assume responsibility, use their heads, develop their own ideas, and implement them in a team. To be honest, I wish that all those who start out at GEMÜ would have the ambition to do their jobs better than I do mine.' This is why the GEMÜ 'Young People Company' was launched in 2006. As part of this initiative, apprentices at GEMÜ assume responsibility for the production and logistics of a selected GEMÜ product and are given the chance to manage projects, individual orders, and special orders independently. The 'Junior Trading Company' initiative launched by the commercial school in Künzelsau, which GEMÜ supports by commissioning the school with small projects, is based on a similar principle.

As long as they successfully complete their apprenticeship, GEMÜ hires all of its apprentices. Particularly talented and hard-working young people are given special support in the form of continual professional training or are given time off work to improve their qualifications. 'Some of the people who now work in our design department started out at GEMÜ as apprentices,' says Müller. 'I prefer to invest in the continual professional training of people whose ability and performance I know rather than buying in specialists from outside.'

The Heilbronn-Franken Chamber of Industry and Commerce and the University of Applied Science in Heilbronn have joined forces to come up with a new degree course known as the 'electrical engineering co-operative degree course'. This four-and-a-half year course combines practical training with courses at the vocational school and the university of applied science. It takes 18 months for students to qualify as an electrician for equipment

Young people studying precision mechanics at the State School of Precision Engineering in Schwenningen, which Fritz Müller supports by getting students to work on real development tasks

and systems. They can then build on this by completing a further three-year course at the Reinhold Würth School (RWH), the Künzelsau branch of the University of Applied Science in Heilbronn, to become qualified engineers or to gain a Bachelor's degree. Since the year 2000, GEMÜ has taken on one applicant for this co-operative degree course every year. He or she receives financial support from GEMÜ for the duration of the degree course and in return agrees to complete his/her practical semester at the company.

The Fritz Müller Foundation also funds scholarships for students at the Reinhold Würth School. Scholarship-holders – most of whom are in the field of electrical engineering or commercial engineering – receive funding for a maximum of four semesters. The intention is that scholarship holders maintain the closest possible contact to GEMÜ. GEMÜ ensures these close links by providing a fixed contact at the company for each scholarship holder. In return, scholarship holders, among other things, complete their practical semester at the company. Through the Fritz Müller Foundation and in conjunction with the district savings bank in Künzelsau, GEMÜ also financially supports the business management degree course offered by the Reinhold Würth School. Moreover, since the end of the 1990s, GEMÜ employees have been working as lecturers at this school and at the University of Applied Sciences in Sigmaringen. GEMÜ regularly offers students practical semesters at the company, topics for dissertations, and student jobs; it supports on request projects being completed by students at the university of applied sciences, and takes part in the so-called 'Think-Ing', an information event organised by the University of Applied Sciences Heilbronn where companies in the region inform first year students about engineering professions and offer activities in which the students can get involved.

GEMÜ's support for institutes of education is not, however, restricted to the Reinhold Würth School. As the founder and president of the association for the promotion of the commercial school in Künzelsau, Fritz Müller supported the school by donating equipment such

Practical training is given top priority at GEMÜ

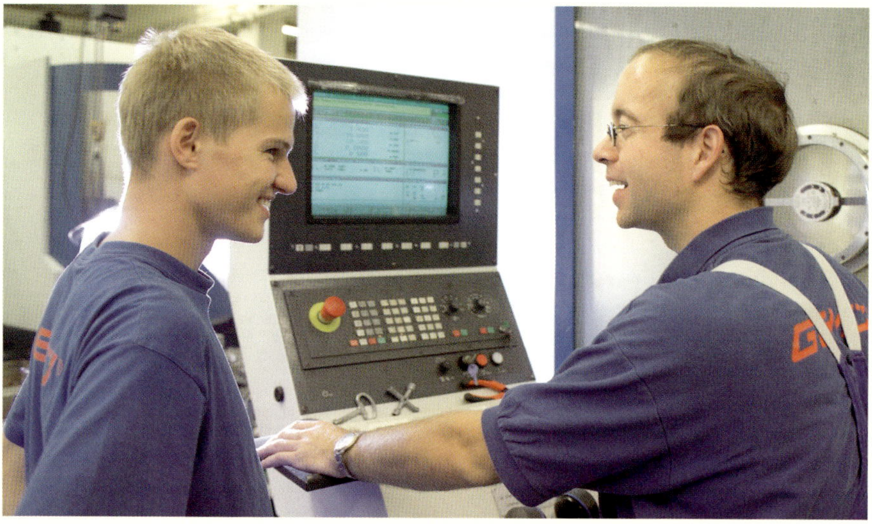

as laptops and a CNC machine. He also promotes the students' projects via GEMÜ. Practical project work is also the main focus of the support Fritz Müller gives to the school he himself attended, the State School of Precision Engineering in Schwenningen. GEMÜ regularly commissions the school to carry out development work. The students in Schwenningen are given a task on which the development department at GEMÜ is working at the same time. Fritz Müller explains the principle behind this procedure: 'What interests me is how people who are not part of the company work on these tasks, what they come up with, and how they come up with the ideas. I also consider it very important that students don't just work on practice pieces, but directly on real projects that are actually being tackled by a real company. I think that the only way to get people excited about their profession in the long term is to show them that their ideas can be directly put into practice and that their work is having an effect.'

As far as Fritz Müller is concerned, however, training and continuing professional development are not only for the young. 'The market is changing so rapidly at present that it is important for everyone to keep enhancing their qualifications in order to stay on the ball in their professions. But what is even more important, I think, is that you remain flexible through learning. Of course, you can be satisfied with what you once learned, but that quickly becomes boring and your work becomes a routine and nothing more. A company like GEMÜ, however, needs people who are interested and awake, who like to start new projects and see them through to the end.' This is why GEMÜ works hand in hand with the KocherJagst Innovation Region initiative and the Heilbronn-Franken Chamber of Industry and Commerce to offer older and lesser qualified job-seekers a two-year training course that qualifies them to become machine and plant operators. Furthermore, since early 2007, any member of GEMÜ staff can improve their language and computer skills by taking courses at the GEMÜ academy or by taking part in internal information events. 'We are delighted whenever anyone is willing to improve their qualifications, hone their skills, assume responsibility, and set things in motion,' says Müller.

CHAPTER 10

EXPANSION INSTEAD OF STANDSTILL: HIGH PURITY, EXPANSION INTO ASIA, AND THE RUNAWAY SUCCESS OF ASEPTIC PRODUCTS (1991–98)

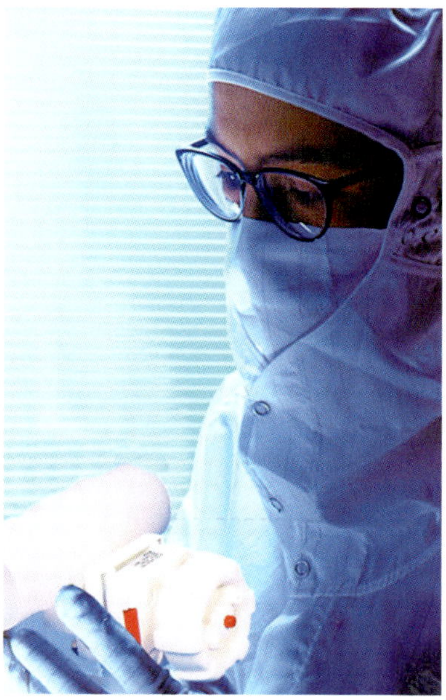

Clean room production

As part of its drive to promote vocational training, GEMÜ began co-operating with eminent German research institutes in the late 1980s. Working with scientists at the Fraunhofer Institute in Stuttgart, for example, the company developed IPA flow controllers for microchip production plants in 1989.

In fact the co-operation with the Fraunhofer Institute went so well that the scientists in Stuttgart encouraged the team in Criesbach to take part in the tendering procedure for the EU research project JESSI (Joint European Submicron Silicon Initiative). The objective of this multi-billion pound research initiative was to create the scientific and technical infrastructure needed for the development and production of high-performance microchips in Europe, thereby reducing the dominance of the USA and Japan in this sector. Microchips are manufactured in 'clean rooms' where the air is kept virtually dust free by systems of filters and airlocks. Because semiconductors are highly sensitive and can be damaged by even the tiniest dust particles or leached trace contamination, clean rooms may only be entered by those wearing special protective clothing. Moreover, the production plants they contain should neither react to aggressive media nor emit particles.

Thanks to its decades of experience in the field of corrosion-resistant plastic valves and its development co-operation with the Fraunhofer Institute, GEMÜ was the only valve manufacturer in Europe to qualify for the JESSI project in 1990. Its involvement in the EU initiative was to prove fruitful in more ways than one. During the six-year funding phase, GEMÜ was able to advance in a targeted manner the development of plastic diaphragm valves and plastic components for use in clean rooms. Through its involvement in this joint research project, GEMÜ made contact with a number of world-leading European electronics companies. Ultimately, GEMÜ's participation in JESSI helped the company become one of the world's top manufacturers of valves and fittings for clean room applications. Because such technology is not only needed for chip production, but also for biotechnology and food production and in the pharmaceutical industry, JESSI put GEMÜ in a position to establish links with companies in these international growth sectors at an early stage. 'Fundamental research and the development of new products are always very costly at the outset,' says Müller. 'But any company that believes it can focus entirely on production instead of investing in these areas will very quickly be left behind. This is exactly why I have always invested heavily in development.'

Innovation proved particularly valuable during the years of recession that were to follow. Increasing globalisation and the advent of digital technology and microelectronics in automated production processes revealed considerable structural weaknesses in west German industry. At the same time, it became increasingly clear that the cost of German reunification (which came about in 1990) and the expense of turning eastern Germany into the 'blossoming landscapes' promised by Chancellor Helmut Kohl would be much higher than initially anticipated in the early euphoric days after the fall of the Berlin Wall.

The Kocher Valley was also hit by the recession. After decades of full employment, numerous companies in the region could only avoid redundancies by putting their staff on short-time work. In 1992, for example, Reisser, an old established screw manufacturer in Ingelfingen that had once employed Marie Müller on the production line and where Fritz Müller's brother, Kurt, spent 30 years working in the commercial department, filed for bankruptcy. Things had changed in another respect too: companies were no longer able to hire all the apprentices they had trained.

Although Fritz Müller was concerned about the way things were developing in the region, he knew that his company was strong enough to withstand the crisis. 'It was clear at the end of the 1980s that a downturn would have to set in at some stage. This is exactly why I invested in the modernisation and extension of the production facility at that time. Instead of hiring more people, we at GEMÜ focussed on improving internal processes so that we could increase productivity,' he says. 'This meant that we were ready for the

Assembling GEMÜ High Purity valves in the clean room (Rotkreuz, Switzerland)

downturn when it arrived. We had the most modern technology available and were in a position to throttle our investment for a period of two to three years. We were also able to hire all our apprentices and didn't have to lay anyone off. No-one was put on short-time work at GEMÜ.'

Confident about his company's strength, Fritz Müller maintained GEMÜ's high quality standards throughout this difficult economic period. He drummed it into his staff that 'as leader in our market niche, we have to refine our expertise as much as possible.' In line with this principle, GEMÜ launched a number of new products for special applications in the early 1990s. In 1990, for example, the company launched its Clean System range, one of its very first valve and fitting ranges for the High Purity (HP) sector. These products were manufactured, packaged, and stored in a specially constructed clean room in GEMÜ's Swiss production facility in Rotkreuz.

To this day, GEMÜ HP valves and fittings are made of PFA, a fluoroplastic that is more commonly known as Teflon®. Thanks to its universal resistance to virtually all aggressive chemicals, its high basic purity, and extremely smooth surface, PFA is ideal for use in all applications involving high-purity media.

Because PFA is only capable of withstanding limited mechanical loads and can, therefore, only be used at low operating pressures and temperatures, conventional HP valves were generally reinforced by a stainless steel casing. This, however, made them too heavy for plants with polypropylene (PP) components such as those that were increasingly being used in the HP sector. Moreover, the stainless steel casings were susceptible to corrosion from external elements and did not meet the strict purity requirements of the HP sector.

The developers at GEMÜ replaced the stainless steel casing with an outliner of PVDF, a lightweight, corrosion-resistant fluoroplastic that met the aforementioned purity requirements. For applications characterised by large nominal widths, the PVDF was reinforced with embedded carbon fibres. These fibres turned the PVDF outliners black in colour, making

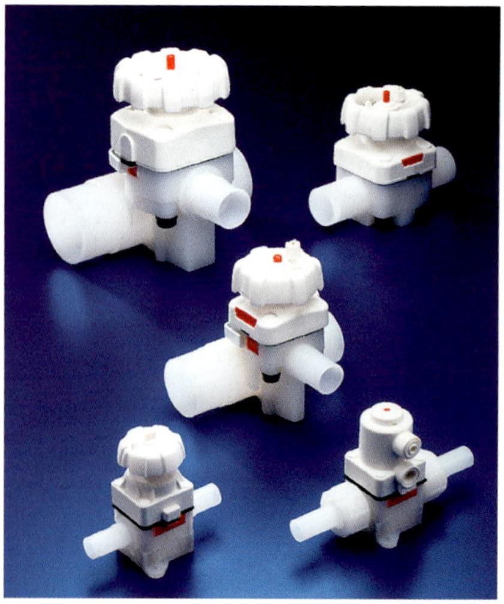

GEMÜ High Purity valves

GEMÜ High Purity valves and fittings in operation (Infineon)

them stand out from the generally colourless PP pipes in the plant. In short, these valves and final controlling elements were easy to identify. The 'clean system' valves were operated either manually or remotely, or were motorised, and as elements in GEMÜ's modular component system, could be varied or replaced at any time. Fritz Müller explains: 'The intention was that customers who could initially only afford a small, manually operated plant, would be able to automate and expand easily as the necessary funds became available.'

GEMÜ's Clean System valves were connected by means of conventional welded joints and hose clamp joints. They could also be integrated into 'clamp fitting systems' in which the parts of the plant could be connected with the help of clamping rings. This connection was absolutely tight and free of deadlegs. Unlike conventional valves, the threaded spigots for the valve bodies in the Clean System range were already integrated into the design, regardless of the mode of connection used. This meant that GEMÜ valves and fittings could be connected to the pipe directly without having to use connectors, which had previously been required, and made it possible to build much more compact plant.

The square design of the piston actuator, which GEMÜ launched at the world's leading trade fair for process automation, INTERKAMA, in 1992, also saved up to 30 per cent more space. Having spent months mulling over the problem of how to reduce the size of the piston actuator, Fritz Müller came up with the idea behind this design – the first of its kind in the world – during his 1991 summer holiday. 'In order to allow the piston to build up the necessary pull, the actuators have to be of a specific size for purely physical reasons. Try as I might, there was just no way of getting around this fact. I had reached a dead end,' recalls Müller. 'Then one day while swimming in Lake Lucerne, I had a brainwave: if there was no way of improving the mechanics, perhaps the solution lay in altering the outer shape! It suddenly occurred to me that there was no law that said actuators had to be cylindrical. If we could make the housing square-shaped, or even rectangular, the actuators could be moved much closer together, making the plant much more compact. I knew at once: that was it! I had made the breakthrough! I swam back to the shore, ran to the nearest telephone, and phoned the development department, giving them instructions to get cracking on my new idea. A year later, we launched the first rectangular piston actuators on the market.'

The piston valves were fitted with a brand new position indicator as standard. This indicator noted even the slightest change in the stroke of the control plunger. The development of this position indicator was in line with GEMÜ's consistent development of its control and measurement ranges in the first half of the 1990s. In 1991, the company launched the GEMÜ 3000, a microprocessor-controlled totalizer that was used, among other things, in automatic filling plants. A year later this was followed by the GEMÜ 3020, a flow transmitter/turbine that was ideal for use in applications featuring aggressive media. In 1993, GEMÜ added measurement equipment with a nominal width of 40 DN to its proven GEMÜ 800 flowmeter range, which had been on the market since the 1970s.

As a specialist for process engineering and feedstocks, GEMÜ not only had a variety of measurement and control equipment and fittings in its range in the early 1990s, but also valves based on all functional principles required by customers, all materials from the fields of plastic and metal, and all kinds of actuators. These components were part of GEMÜ's modular system, elements of which could be combined in just under 300,000 different ways. The scope and depth of this range meant that the Criesbach-based company saw out the economic downturn of 1992–94 relatively unscathed. 'We were able to compensate for slumps in certain areas by simply shifting to other areas,' recalls Müller.

Thanks to GEMÜ's rectangular piston valves, plant manufacturers could save up to 30 per cent space

Spreading the risk did not, however, guarantee protection against force majeure. Following days of heavy rainfall shortly before Christmas 1993, the river Kocher rose to its highest level in 100 years within the space of a few hours. The river burst its banks and flooded the entire industrial heartland of Criesbach. While part of the workforce scrambled to move the company's most important plant and stock from the basement of the building to a safer, drier location, others worked feverishly outside to keep the swollen river at bay using sandbags and hastily acquired pumps. But their efforts were in vain: the muddy waters of the flood swamped the basement and ground floor of the GEMÜ factory building, submerging everything that had not been brought to safety in time.

Employees joined forces and tried to keep the flood waters at bay

Despite the desperate situation in which they found themselves, Fritz Müller and his 350-strong workforce in Criesbach refused to let themselves be discouraged. Together they had quite literally risen from the ashes of the company after the fire twelve years previously; they would now pull together and overcome the flood. As soon as the waters receded, the GEMÜ team started to clean up. The flooded basements were pumped dry and machines and stock were checked. Once this was done, the focus switched to renovating the building as quickly as possible. By the end of January 1994, only a few weeks after the flood, GEMÜ relaunched production in Criesbach. The experience of having pulled together and overcome a serious threat to the company's existence brought the staff even closer together.

Sadly, their efforts were in vain: the swollen river swamped the basement and parts of the ground floor, causing millions of Marks worth of damage

The sound support the company experienced at its headquarters in Criesbach and the continuing success of GEMÜ's subsidiaries abroad encouraged Fritz Müller to push ahead with the international expansion of his company. He began by focussing on France. Although GEMÜ had established a small sales company with a little warehouse in Paris in the early 1980s, the French subsidiary's trade had stagnated at a low level. Müller remembers: 'Whenever I drew the attention of my managing director in Paris to the poor figures, I got excuse after excuse. Sometimes it was the fault of the long French summer holidays, sometimes the fault of a strike. He always had some excuse.' In the autumn of 1994, Fritz Müller laid off his managing director in Paris and decided to run GEMÜ's operations on the French-language market from Duppigheim in Alsace. Within the space of only a few months, an ultra-modern warehouse was erected in Duppigheim near Strasbourg under the watchful eye of Hermann Walter, a long-serving GEMÜ employee, who liaised closely with GEMÜ's commercial director, Eckhard Noller. Since the spring of 1995, all of GEMÜ's European subsidiaries have been supplied from this warehouse. Duppigheim is also the company's sales headquarters for France and other French-speaking countries in Europe. 'The French, and the Parisians in particular,' says Müller with irritation, 'consider the people of Alsace to be uncultivated country bumpkins. But after all the problems I experienced with the Paris office down through the years, it was a relief for me, as a German, to work in Alsace: I not only had fewer language problems, but found that their mentality was closer to the German mentality than any other in France. That being said, the people of Alsace are French enough to be able to do good business with their compatriots.' The change in location was a good move for GEMÜ s.a.r.l. 'Ever since we began running our French operations from Duppigheim, we have been in the black there,' relates a satisfied Müller.

In 1995, GEMÜ's warehouse in Warrington, England, was extended after the company concluded a contract with a leading British pharmaceutical company for the supply of HP diaphragm

The premises of GEMÜ s.a.r.l. in Duppigheim, France

valves. This contract involved the guarantee that GEMÜ would be able to supply any of the customer's plant with spares within 24 hours. While the extension of the warehouse made this promise easier to keep, it was not always so. GEMÜ's British customers always demanded a rapid service even before the extension of the warehouse. On one occasion, for example, the Managing Director of GEMÜ UK flew to Ingelfingen in person to get an urgently needed spare for a customer and then flew back with the part and drove to the customer to deliver it personally the same day. On another, when a major biotechnology company in the south of England needed help repairing a plant (the company's design engineers had installed unsuitable valves and fittings), a German team of technicians arrived on site to lend a hand within hours.

Fritz Müller wanted to be able to offer his overseas customers the same kind of service and support. 'Although we had built up a global network of sales companies by the mid 1990s, I knew that in order to assure designers, plant manufacturers, and operators that we would always be able to deliver the spares and provide the technical support they needed in the shortest possible time frame, we would have to actually be on these markets – ideally with our own production plants in order to avoid the expense of transportation and storage', says Müller.

This meant that GEMÜ would have to further its international expansion. This time, however, Fritz Müller did not look West as he did in the 1970s, but East to Asia. The first boom in the low-wage tiger economies of Asia was already over by the middle of the 1990s. Wages in Taiwan, Singapore, and South Korea had risen and the people in these countries were increasingly aiming for a more western standard of living. In order to ensure the economic growth necessary for such a standard of living, the Asian states had to build up a modern industrial sector as quickly as possible – just like Japan had done 20 years previously. This is why in the second half of the 1990s, these countries came to the attention of GEMÜ and its customers. After all, modern industry needs automated plants and, therefore, valves, fittings, and measurement equipment. 'The Asian market had enormous potential and was virtually a completely new market for us,' recalls Müller. And so, in 1995, GEMÜ opened its first office in Asia in Singapore's 'German Centre'. Two years later, the Criesbach-based company opened an office in Shanghai, its first foothold on the Chinese market, which was at the time the world's second largest industrial growth market after Brazil.

While GEMÜ was reinforcing its network of foreign branches, more production space was needed back home at company headquarters in Criesbach. In August 1996, Fritz Müller laid the foundation stone for a 4,000-square metre extension. At the same time, the team at GEMÜ's headquarters were working feverishly on new products. Since the mid 1990s, GEMÜ's range of products had included valves and measurement equipment for all available field bus systems on the market. This communication system had been in widespread use in automated industrial processes since the mid 1980s. The invention of the field bus meant that the bulky, high-maintenance parallel cabling of a number of different field devices in industrial plants could be replaced by compact, inexpensive serial connections. With a field bus system, a single cable (the so-called 'bus cable') links up all levels of the system – from field devices such as sensors, final controlling elements, and actuators to the control unit.

The use of programmable logic controllers (PLCs) – microprocessor-controlled compact assemblies for the control of pre-programmed processes – in automated production had been spreading since the 1980s. In the early 1990s, field bus and PLC technologies were used to create Local Operating Networks (LONs). In a LON, the central PLC control system is replaced by decentralised, freely programmable field device controls. The individual actuators and sensors became autonomous nodes that could be connected at will and could exchange information directly with each other in order to trigger actions in any participant within the LON. LONs can also be connected to one another via intranet and Internet, thereby allowing a company to integrate a number of locations into a single network.

Fritz Müller and works manager Hermann Walter lay the foundation stone for the extension (1996)

This technology was also adopted immediately by GEMÜ's developers. Since 1995, GEMÜ's modular system has also included LON-compatible valves and measurement equipment. In 1996, for example, GEMÜ valves and fittings were installed in a LON used by a textile manufacturer based in the German city of Schwäbisch Hall and used to control dyeing processes in its production unit in China via Internet.

Increasingly closer networking also became a feature of internal organisation at GEMÜ. Since the early 1990s, a manufacturing requirements planning (MRP) system has linked up GEMÜ's various production and organisational units and its most important international locations. A management information system also provides managers with daily updates on the most important business figures. 'With thousands of different product versions, batches of between 10 and 100,000 units, and delivery times that can be as short as under a week for special orders, the

GEMÜ field bus valves and fittings

only way to ensure that we work economically and meet our deadlines is to harmonise all internal processes at the various locations – from development through production and right up to the despatch of the finished product, including all the associated commercial aspects,' explains works manager Hermann Walter, who was responsible for the introduction of the MRP system at GEMÜ and is to this day responsible for its upkeep. In order to make sure that the software met the company's practical requirements, as many as possible of the staff's requests and suggestions were taken into account in the run-up to the launch of the MRP system. There was also a theoretical side to the implementation of the system: Matthias Fick, the son of Fritz Müller's childhood friend and long-serving employee, Emil Fick, made these optimisation processes at GEMÜ the subject of his dissertation for his degree in commercial engineering. Because of its highly practical approach, Fick's dissertation was awarded the University Prize of the Chamber of Industry and Commerce in East Württemberg in 1994.

It could well be that this permanent quest for process optimisation and intelligent networking paved the way for the development of stainless steel multi-port valves. It was this ground-breaking new development that made GEMÜ one of the world's leading manufacturers of aseptic valves and fittings in 1997, a position it has maintained ever since. The engineer Jochen Hofmann, son of Fritz Müller's long-serving employee Manfred Hofmann and head of the aseptic technology unit for more than 15 years, remembers the advent of this success story: 'In the early 1990s, I was involved in GEMÜ's ISO certification process as a quality assurance officer. In 1992, Fritz Müller offered me the job of head of the biotechnology unit (now the aseptic technology department). He put a designer and two

Modern production techniques at GEMÜ headquarters in Criesbach

sales reps at my disposal and told me that I could do what I liked as long as I made GEMÜ no. 1 in the biotechnology sector. "See to it!" he said,' says Hofmann, laughing.

This was no easy task considering that unlike its products for the high purity sector, GEMÜ's range of products for the biotechnology sector in the early 1990s comprised virtually only standard valves and fittings. Originally developed for sewage treatment, these GEMÜ valves were not suitable for use in aseptic applications. Nor did the product documentation reach the high standards of the aseptic technology sector. 'Expectations in the pharmaceutical industry, for example, are much higher than in other sectors,' explains Hofmann. 'We had to do a complete rethink in order to be able to supply these customers.'

Together with his team, Hofmann visited plant manufacturers and operators and familiarised himself with their production requirements and quality standards. During this period, the team realised that networking and process optimisation were also major features of this industry. Many aseptic technology applications required complex ring mains in which numerous different valves and fittings were connected using pipes. At the time, these systems were built using T fittings, reducers, and welded-on standard valves. However, as a result of the many welds, these systems required frequent repairs. Moreover, deadlegs often occurred around T fittings and sometimes in the valve bodies themselves, making them virtually useless for applications in the sensitive pharmaceutical sector. In short, the pharmaceutical industry was crying out for deadleg-free ring mains with as few welds as possible. This was the challenge Jochen Hofmann and his team brought back to Criesbach.

The GEMÜ developers set to work. They reduced the number of welds by integrating the connecting piece into the valve body itself (similar to the design of GEMÜ's HP valves). They avoided deadlegs in the valves and fittings by only using self-cleaning

Compact and without deadlegs: GEMÜ's multi-port valves for aseptic technology applications, on its own (X form shown here) and in series

deadleg-free diaphragm valves. The only remaining challenge was the elimination of deadlegs in the T fittings. Faced with this tricky task, the developers asked themselves whether these fittings were really necessary at all. 2/3-port T valve bodies had been in use in the valve sector for quite some time. Would it not be possible to design a multi-port valve body, thereby integrating the connection into the fitting?

In 1995, GEMÜ launched its first multi-port T valve body with integral connector on the market. This was soon followed by X, Y, and H valve bodies. These stainless steel milled valves allowed not only for the construction of extremely compact ring main systems for the aseptic industry, but also for the design of complex networks and – by welding a number of valve bodies together – the creation of entire valve nodes with complex transit paths.

With these new T, Y, X, and H multi-port valve bodies, GEMÜ went from strength to strength on the market for aseptic valves and fittings. But customers quickly began making new demands: they needed valve bodies with even more ports and increasingly complex designs (geometries). From the outside, the valves and fittings had to be as fully enclosed as possible. Ideally speaking, they should have had no connecting points at all. These demands meant that GEMÜ's designers were faced with an almost impossible task: it was generally held that the only way to create valves and fittings with the required high number of ports and such highly complex geometries was to combine individual valve bodies. In short, a certain degree of bulkiness and a minimum number of joints were unavoidable.

'The old design principle was out-dated,' says Fritz Müller. 'If we wanted to make progress, we had to approach the matter from a completely new angle. The solution lay not in the mechanics, but in the shape.' Jochen Hofmann explains: 'At some point, we realised that it made no sense to mill lots of individual valve bodies out of steel only to weld them back together again to form polygonal, bulky structures. Seeing as everyone was giving much thought to optimising interfaces and clearly structured user interfaces, it made sense to machine compact, interface-optimised components out of a solid block of metal instead of patching a number of different elements together. We had the production plants and qualified staff we needed to swiftly turn this idea into a product that would meet the customers' expectations.'

GEMÜ launched the very first multi-port valves on the market in the spring of 1997. At the time, even Fritz Müller and Jochen Hofmann had no idea of the potential of this new product, especially as dedicated magazines had not shown a great interest in the new products, and the marketing of these valves had as a result been rather subdued. The breakthrough came when GEMÜ exhibited the multi-port valves at ACHEMA, one of the world's leading biotechnology trade fairs, in August 1997. After the trade fair, GEMÜ's incoming orders for aseptic valves tripled. In order to be able to meet delivery deadlines, staff worked overtime and new workers were hired. In the course of only a few weeks, Fritz Müller invested just under DM 10 million in the development of the company's aseptic technology unit. New machines were purchased and the extension project that had got underway twelve months previously was extended again by a further 2,000 square metres. 'When you have a new product that is selling like hot cakes, you have to react quickly and invest in order to stay ahead of the competition, otherwise you are giving your competitors the chance to catch up on you,' he explains. And he was right; the investment was worthwhile. Before the year was out, GEMÜ was European market leader for aseptic valves and fittings. Two years later, it was already one of the world's leading companies in this sector. Aseptic valves now accounted for 30 per cent of all products manufactured by GEMÜ and its ten subsidiaries. At the end of the 1990s, GEMÜ employed over 750 people worldwide; in Criesbach alone, 200 new jobs had been created since 1989.

A GEMÜ multi-port valve

Fritz Müller was well aware of his growing responsibility. As far as he was concerned, retirement – which the press speculated about on the occasion of his 60th birthday in January 1999 – was only an option in the medium to long term. Whenever he was questioned on the matter, he would reply: 'The boss cannot just get up one morning and say, "I'm off. You'll have to manage by yourselves from now on!"' On the contrary, he considered it is his responsibility to lead the company successfully into the new millennium and to gradually pave the way for the day he could retire, safe in the knowledge that GEMÜ would be able to go on growing and developing as he wished.

HAVING STRONG LOCAL ROOTS IN A GLOBALIZED WORLD

The more international GEMÜ became, the more frequently Fritz Müller had to travel abroad on business trips. In 1996 alone, for example, he spent 229 days visiting customers and employees in all four corners of the globe. In the course of his regular visits to GEMÜ's Brazilian subsidiary in Curitiba, he repeatedly came face to face with poverty, underdevelopment, and oppression. 'There is a huge amount of poverty in Brazil, but only very, very few poor people actually get any kind of support. It is very difficult for poor children; for poor orphans the situation is utterly disastrous,' says Müller. He is firmly convinced that a company that generates profit from working in a country is also obliged to do something to address the needs of the people in that country. And so, he and his wife became sponsors of a Curitiba home for just under 150 boys aged from 7 to 18 years. Back in Hohenlohe, he asked members of the Rotary Club – of which he himself is a member – for money and donations in kind. The staff at GEMÜ also collected money at Christmas for the children's home in Brazil. With the money they received, the children's home built up a small farm and fish farm as a source of food. Both projects developed so well that the home is now economically independent.

The opening of the home for drug-addicted street children in Curitiba that was launched by GEMÜ's former works manager, Manfred Hofmann, and partly funded by GEMÜ

Both of the homes supported by GEMÜ introduce children to everyday activities – such as preparing meals together, as shown here – at an early stage in order to promote their independence

Encouraged by this success, GEMÜ's former works manager, Manfred Hofmann, got involved in the construction of another

home in Curitiba in 1997. On Hofmann's initiative and with funds provided by GEMÜ, a Church-run home was built outside Curitiba in the rainforest. At present, it is home to approximately 50 so-called 'sniffer kids', homeless children who live on the street and regularly sniff glue and solvents to escape the drudgery of their lives. The GEMÜ-sponsored home gives these children and teenagers, who are seriously addicted to drugs, the chance to kick their addiction and rehabilitate themselves socially.

However, as time goes by, Fritz Müller has less and less time to focus intensively on the countries where GEMÜ has subsidiaries. In the late 1990s, he compared his daily routine while abroad with that of top-class sportsmen and -women. 'Airplane, hotel, competition, next competition ... that's a bit what it's like for me when I am travelling the world as a global player for GEMÜ,' he said with regret in an interview at the time.

This is why it was all the more important for him to switch off in his free time. As time went on, he spent less and less time gliding. Although Fritz Müller is still president of the Kocher Valley Air Sports Club and has donated and presented a variety of prizes that seek to foster the talent of gifted glider pilots in the region since the year 2000, he himself rarely finds the time to set off from his local airfield or to play an active role in club life. In view of the numerous business meetings to which he flies himself in his own airplane, he rarely has the urge to take to the skies again when he is in Ingelfingen.

Be that as it may, Fritz Müller is certainly not averse to streamlined means of transport when on the ground. In 1987, he treated himself to a Ferrari 328 GTS for his 48th birthday. Two years later he bought a Ferrari Testarossa. It was not the last model from Maranello he was to buy. 'Initially, it tickled me to tell everyone that I had grown tired of being overtaken by more powerful cars on the motorway,' says the car enthusiast, 'but it had much less to do with the actual horsepower of the car; I just find Ferraris beautiful. For me, these cars are the perfect combination of engineering art and design.'

This synthesis of form and function can be found in a number of vehicles in the entrepreneur's extensive collection, which now comprises over 140

The first car in Fritz Müller's collection: a Ferrari 328 GTS

vintage cars and motorcycles. 'I really started out collecting cars in memory of my father,' relates Müller. 'Before the war, he had two Mercedes-Benz Stuttgart 260s, one a sedan and the other a convertible. Both cars were requisitioned during the war; we never saw them again. But my mother often spoke of them and there were, of course, photos of them. So when I began collecting cars, it occurred to me how nice it would be to have one of these cars – a kind of souvenir of my father.'

After searching for several months, Fritz Müller found the convertible version of the model he was looking for in Berlin. The next vintage car in the Müller collection, an old BMW Isetta, which Ingrid Müller gave her husband for his 50th birthday in 1989, was a reference to a different story, one that involved both husband and wife: in his first months at Herion, Fritz Müller drove to work every day in a striking Isetta, which was more

widely known in Germany as the 'asphalt bubble' or 'kissing car'. 'Once I had the Isetta, I just had to get a Mercedes 180, the car that so enraged my superiors at Herion,' says Müller with a grin. His passion for collecting was burning brightly.

What fascinates Fritz Müller the most about vintage cars is the unusual technical solutions they contain. For him, these innovations symbolise the spirit of invention and sometimes the tragic reality of failure. In his office, for example, stands a faithful copy of a motorcycle dating from the year 1894. This copy was partially built on GEMÜ machines. 'For me, this motorcycle is a perfect example of what happens when you rigidly stick to out-dated solutions instead of searching for new approaches,' he explains. The designers of the motorcycle in question, Heinrich Hildebrand and Alois Wolfmüller of Munich, were the first to manufacture motorcycles in series production. They even secured themselves the right to the exclusive use of the word 'Motorrad' (motorcycle). Initially, their motorcycles sold well. However, the Hildebrand & Wolfmüller motorcycle had one decisive design error: the power from the engine was transmitted directly to the rear wheel by means of two connecting rods, as was the case with locomotives. Transmission or idle did not feature in this design at all. Once in motion, the only way to stop the motorcycle was to exert a large amount of pressure on the front wheel by means of its wooden brake. The braking distance was correspondingly long and, frequently, the only way

Fritz Müller bought this Mercedes-Benz Stuttgart 260 convertible in memory of his father

Copy of a Hildebrand & Wolfmüller motorcycle (1894)

for the motorcyclist to avoid an impending collision was to leap off the moving motorcycle. Although Hildebrand and Wolfmüller feverishly tried to get a handle on the problem using increasingly ingenious solutions, the actual design of the system that powered the motorcycle remained unchanged. 'Despite their numerous attempts to improve it, these machines were really nothing more than uncontrollable locomotives on two wheels,' says Müller, 'and after a year in business, Hildebrand and Wolfmüller went bankrupt.'

For Müller, lessons of a very different nature altogether can be learned from the story behind another of his favourite items, a Veritas RS 2000, a German racing car dating from 1947. 'For me, the story of the Veritas is the epitome of entrepreneurial spirit and the art of improvisation in the post-war years,' he says enthusiastically. In 1947, a team of former BMW employees led by the former racing driver and BMW designer Ernst Loofs, transformed an old shed in Messkirch in the Baden region into a small garage and began making racing cars. 'When you think about it, it was a completely lunatic idea,' says Müller. 'The Americans and the British had imposed a ban on motor racing in Germany; neither metal nor spare parts were available; many roads had been destroyed, and no-one but no-one needed racing cars. Anyone with half an ounce of sense would have repaired trucks or jeeps for a living. But not Ernst Loofs. He decided to concentrate on doing what he did best, namely building racing cars. He was convinced that the ban on motor racing would be lifted sooner or later and that when that day came, he would find buyers for his cars. Moreover, he knew a few good designers and mechanics who were just as mad as he was and were willing to join him.'

Because there was no new material available to use in his cars, Loofs recycled old BMW sports cars that his customers gave him to rework or which he obtained from BMW dealers he knew in exchange for butter and bacon. Loofs then upgraded these second-hand cars to racing cars using the simplest of materials and a lot of technical skill. The BMW chassis remained intact, but the engine performance was cranked up from 80 to

115 hp and the heavy original car body was replaced by an aerodynamic aluminium skin supported by a stable, yet lightweight metal grid construction. Georg 'Schorsch' Meier, who won the 1948 German Formula 2 championship in a Veritas, recalled that the whole structure was frequently held together by little more than insulating tape and wire. 'In principle, there wasn't much difference between the Veritas and the soap-box cart that my brother and I built around the same time in Ingelfingen,' says Fritz Müller with a laugh, 'so of course I just couldn't resist when someone offered to sell me a Veritas RS 2000 almost 50 years later.'

Fritz Müller and his co-pilot, Thomas Feierabend, in his Veritas RS 2000 at the start of the mountain race at the Langenburg Historic Rally 2008

Fritz Müller regularly takes part in vintage car rallies such as the annual Langenburg Historic, which is sponsored by GEMÜ, with his Veritas or other old cars in his collection. He has absolutely no qualms about taking part in difficult mountain races in one of his valuable collector's items. 'After all,' he says, 'these cars were built as racing cars. That being said, I am just as happy participating in the touristy side of the rally or just heading out in one of the cars with my wife at the weekend and taking a leisurely drive through the country.'

Despite his planes and vintage cars, the entrepreneur also likes to get around on foot. Says Müller: 'To this day I find that the best way to come up with new ideas is to go for a walk along the river or around the vineyards.' Moreover, as a member of a winemaking family, it is in his nature to want to keep an eye on the vines at regular intervals, especially since he bought some of the best vineyards in Ingelfingen in the early 1990s. 'I come from around here, I love the landscape and, of course, the local wines,' says Müller who himself makes wine in his spare time.

He loves to tell people – foreign business partners included – of the beauty of his native country and the quality of Ingelfingen's wines. It is tradition in the Müller family to invite business partners into the family home high above the Kocher Valley near Niedernhall. The couple moved here in the early 1980s following the fire that destroyed GEMÜ's premises and their flat. Ingrid Müller personally saw to the extension and renovation of this 1960s villa and the landscaping of the extensive gardens surrounding it. One of the first additions to the site was the 'barbeque hut' where Ingrid Müller likes to cook for her husband's guests. Faced with the problem of not being able to entertain larger parties in this hut, the Müllers had the idea in late 2000 of building a tavern-like building at their vineyard at Hoher Berg in Ingelfingen.

Ingelfingen's authorities, however, refused to grant permission for the construction of a private building in the vineyards that would not be used for making wine. Undaunted by this rejection, Fritz Müller and his old friend Max Vogelsang, the head of a Swiss design and construction office

The staves for the barrel were constructed in Max Vogelsang's carpenter shop in Switzerland and put together in the traditional cooper manner to create the 'Ingelfingen Barrel'

for non-standard wood structures, put their heads together and came up with a completely new design, which they presented to the town's authorities as a completely revised submission only a few weeks later.

This new submission was not for a vineyard tavern, but for a building in the form of an oversized wine barrel lying on its side ... and much more besides. The intention was that the 'Barrel', the second largest of its kind in Europe, would – like its bigger sister in Bad Dürkheim in the state of Rhineland Palatinate or the world famous giant cask at Heidelberg Castle – become a tourist attraction. The ground floor of the building would house a wine-making museum where old wine-making equipment and tools could be viewed at all times through large windows. The function room, which would comfortably hold at least 30 people, would be available not only to GEMÜ, but also to any other parties interested in renting it for a function. Open Days would be organised twice a year and the town of Ingelfingen would be allowed to use the 'Barrel' free of charge for a certain number of events every year. Finally, the submission also foresaw the construction of a tourist path leading from Ingelfingen to the 'Barrel' and back again that would provide visitors with information about the history of wine-making in the Kocher Valley.

Max Vogelsang, an aerobatics pilot and aircraft designer, not only planned and built the 'Ingelfingen Barrel', but also designed a two-seater aerobatics plane, the VOTEC 322, for his friend Fritz Müller

The authorities were satisfied and approved the plans. In July 2001, construction of 'Fritz's little barrel' (as the barrel became known in Ingelfingen) got underway. While the foundations were laid and the walls of the wine cellar were built by the Künzelsau-based architectural studio Erich Kalis, Max Vogelsang's carpenter shop produced the 15-m long

INGELFINGERFASS

wooden staves for the barrel structure. These staves were delivered to Ingelfingen in autumn 2001 and put together in the traditional manner practiced by coopers for centuries. Subsequently, the 9-metre high, 18-metre wide self-supporting structure with a volume of over 700 cubic metres was secured like a normal wine barrel by attaching prestressed steel hoops.

Once the inside of the barrel was decorated and the 25 stations of the tourist path had been erected, the 'Ingelfingen Barrel' was opened in August 2002. The opening of the building attracted much attention. Ever since, the wine-making museum has been a popular tourist attraction. Companies and clubs use this unusual location, as do the town of Ingelfingen and the Müllers on a regular basis. On one occasion, they played host at the 'Barrel' to the German women's national fistball team, which is sponsored by GEMÜ, or the voluntary fire service of the town of Ingelfingen. Erika Brauch, the daughter of a wine-maker, and the team from the Schloss Hotel Ingelfingen provide the catering at the 'Barrel' on such occasions.

Once the foremost hotel in the locality, the Schloss Hotel Ingelfingen, which takes up one wing of Castle Ingelfingen, had been becoming increasingly run down ever since the start of Ingelfingen's decline as a spa and cure location in the late 1970s. By the end of the 1980s, the building in the town's old quarter stood empty. The town's authorities spent months searching for a new tenant to run the hotel. When no-one came forward, they eventually turned to the large companies in the region for help. Fritz Müller and two other entrepreneurs agreed to renovate the dilapidated hotel and run it together.

When both his partners pulled out of the operation in the late 1990s, Fritz Müller decided to continue running the three star hotel with its twenty modern rooms himself. 'Although from a business point of view there are more lucrative investments, it was very important for me to be able to offer our customers and, of course, other visitors to Ingelfingen, comfortable accommodation,' says Müller. The hotel, its restaurant, the

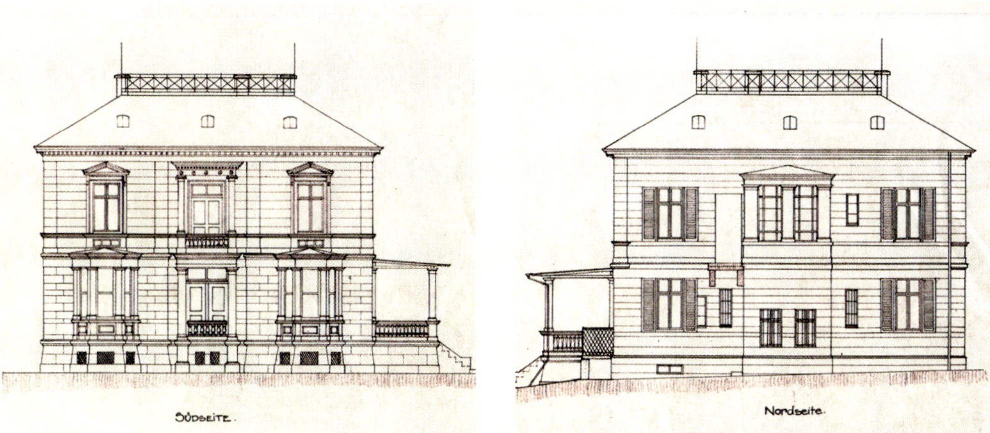

Historic plans of Villa Schüßler in Künzelsau

interior decoration of which was taken care of by Ingrid Müller, and the high-quality regional style food is not only appreciated by hotel guests, but also by discerning local diners.

Now experienced in the renovation of historic buildings, Ingrid Müller joined forces with architect Erich Kalis, who had also been involved in the 'Ingelfingen Barrel' project, to restore Villa Schüßler, a listed building in Künzelsau. This classical-style manufacturer's villa, which had been used by the city of Künzelsau for many years, was no longer in use and was bought by Fritz Müller in 2005. Since the completion of renovation work in summer 2007, this beautiful old building has housed the VIP Präsente and GFM Solutions companies, which are run by the Müller children, Regina Kalis and Gert Müller respectively.

Back home in Ingelfingen, Fritz Müller is personally overseeing the renovation of the historic Gatekeeper's House at what used to be the town's north gate. When the town began looking for people who were willing to support the restoration of the listed eighteenth-century building as part of its project to restore Ingelfingen's historic centre in 2006, Fritz Müller bought the building from the city and undertook to have the Gatekeeper's House and the historic tower clock restored in accordance with requirements for listed buildings. As soon as the restoration is finished, Fritz Müller would like to turn the building,

Once renovated, the Gatekeeper's House in the historic town centre of Ingelfingen will be used as a museum of precision engineering and history of Ingelfingen

which stands opposite Castle Ingelfingen, into a little museum that will be open to the public. 'It pains me to see an old building that I have cherished since my childhood becoming run down or being demolished,' says Müller. 'Although I am very much in favour of innovation, that doesn't mean that I like to see old things simply being thrown away or standing empty. On the contrary, the most sustainable new ideas are those that combine the old with the new, or where an old design principle is reinterpreted in a new way. This is why I am involved in the restoration and re-use of listed buildings in Ingelfingen. You have to fill the past with life so that it can become the starting point for something new.'

CHAPTER 11

NEW BEGINNINGS: INTERNATIONAL CHALLENGES AND THE GEMÜ DOME (1998–2008)

Fritz and Ingrid Müller at the opening of GEMÜ's Austrian subsidiary in autumn 2000

Fritz Müller knew exactly the challenges faced by GEMÜ in the early years of the new millennium: the company would have to develop more innovative products, especially those known as 'system solutions', while at the same time continually enlarging its international network. No sooner had the extension of the company building in Criesbach been completed in 1998 than construction of the new premises of the Swedish GEMÜ subsidiary began in Lindome near Gothenberg. After a very subdued start to business after the establishment of GEMÜ Armatur AB in 1981, the subsidiary's turnover tripled in the 1990s. The sales company not only supplies GEMÜ's Swedish customers, but is also responsible for the entire Scandinavian market. GEMÜ GmbH, GEMÜ's Austrian subsidiary, which was established in October 2000, works a similarly large market. From its base in Guntramsdorf near Vienna, this sales company run all of GEMÜ's operations in Eastern and South-Eastern Europe.

Despite these European developments, GEMÜ had been focussing its international efforts on markets outside Europe since the year 2000.

GEMÜ's Brazilian subsidiary continued its strong development and was responsible for a considerable part of the company's export turnover. Some 75 per cent of the metal valves, diaphragms, and accessories manufactured in Curitiba are sold in Latin America; the remainder in North America and South Africa, where GEMÜ opened its own agency in late 2007. The success of GEMÜ Brazil motivated Fritz Müller to further the globalization of his company. Before sending staff out on this mission, he would tell them: 'There is still a lot of uncharted territory on GEMÜ's world map. We now have the chance to discover these territories bit by bit and build up a presence in them.'

GEMÜ's agencies in China

Opening the headquarters of GEMÜ's Chinese operations in Shanghai (2006)

The initial focus of this international drive was in South-East Asia. The Shanghai branch established in 1997 had developed so well that Fritz Müller decided the time had come to set up a production plant there. In November 2000, the newly established subsidiary, GEMÜ Valves (Shanghai) Co. Ltd., opened its first small production plant. In the early days, four Chinese employees, who had been trained in Criesbach, manufactured mainly butterfly valves for the Chinese and East Asian markets.

A year later, GEMÜ opened a sales office in Beijing and in 2002, agencies in Guangzhou, Shenyang, and Chengdu in western China followed. In 2003, GEMÜ Valves (Shanghai) Co Ltd. was already responsible for

GEMÜ®
盖米阀门(上海)有限公
GEMÜ Valves (Shanghai) Co.,

GEMÜ®
德国盖米阀门.测量及控制系统
新起点 新挑战
New Start New Challenge

posting 15 per cent of GEMÜ's total export turnover. In the years that followed, this turnover regularly grew by over 10 per cent, and in 2006, the headquarters of the company's Chinese operations in Shanghai were extended. The new building, which was modelled on European GEMÜ subsidiaries, covered 3,500 square metres and housed administrative offices, production facilities, and storage space. It is now the logistics hub for GEMÜ's activities in South-East Asia.

Opening the headquarters of GEMÜ's Chinese operations in Shanghai (2006)

GEMÜ not only has numerous sales partnerships in the region, it also has several agencies. In 2003, for example, it opened an agency in Taiwan and established a subsidiary in Malaysia. This was followed in 2006 by sales offices in Indonesia and India. 'The markets of South-East Asia have huge growth potential,' says Fritz Müller. 'On the one hand, modern industries are being built up at high speed, especially in the semi-conductor and pharmaceutical industries. This is an ideal market for our HP and sterile products. On the other hand, the problems of environmental pollution are growing and resources are diminishing. Singapore, for example, depends largely on Malaysia for its drinking water. Because this was repeatedly a bone of contention between the two countries, Singapore is now building massive sea water desalination plants that are fitted with GEMÜ valves and controls.'

Water treatment plant fitted with GEMÜ valves and controls

GEMÜ's swift growth in Asia was due to a number of factors, not least of which was the fact that staff made a concerted effort to adjust to the mentality of their business partners. 'Regular visits are considered a sign of respect in Asia. Questions have to be answered quickly and pronounced politeness is very, very important,' explains Müller. This is why GEMÜ managers from Germany visit the company's Asian branches and customers as often as possible. Moreover, an

English-speaking 'Asian team' is on hand in Criesbach to deal with questions and problems – from customers and subsidiaries alike – by telephone and e-mail from 6.30 a.m. Monday to Friday.

Fritz Müller also has great expectations regarding the development of the Russian market. GEMÜ's dealings with Russia first began in 1986. InterApp, which had been selling GEMÜ products behind the iron curtain for almost twenty years , arranged a meeting with contacts in the Soviet Ministry of Economic Affairs. In May 1987, after several weeks of negotiations – during which both sides tested their ability to hold drink – GEMÜ was given permission to construct a plant that would manufacture its valves under licence in Tambov on the River Don, 500 km south of Moscow. Wolfgang Schütte, a long-serving GEMÜ employee, oversaw the construction of the plant.

It would seem that production initially progressed very well after the plant was handed over to the Russians. However, contact with the plant in Tambov ceased after the collapse of the Soviet Union. When Fritz Müller was later offered the plant for sale, he and Wolfgang Schütte flew to Tambov. They were appalled by what they found. 'Everything was run down,' he recalls. 'The production facility was just one big Potemkin village! They had put a few people at the machines and told them to pretend they were making valves. Even if they had known how to make valves, they couldn't possibly have done so using those machines; there was hardly anything left of them. Anything that hadn't been nailed down had been dismantled; electric motors were missing; even the simplest tools were gone.'

Signing the contract in Moscow (1987): Fritz Müller recalls that the cans of cola were already opened – and filled with vodka – when brought to the table

Horrified by what he saw in Tambov, Müller decided not to further his involvement in the Russian market. However, in view of the fact that the country has recovered in recent years and income from oil and gas production is being invested in industry, Fritz Müller now sees opportunities again for GEMÜ in Russia. 'The Russians want to improve to western standards as quickly as possible and they have the experts they need to do so. This is a massive growth market for GEMÜ,' says Müller. Reflecting the significance of the market for the company, GEMÜ opened its first sales office in Moscow in spring 2008. With the support of a Russian-speaking customer centre in Criesbach, this office is working the markets in Russia, Ukraine, Belarus, and Kazakhstan. A warehouse for spares and wearing parts is currently being planned.

Even during this period of international expansion, GEMÜ continued churning out technical innovations. At ACHEMA 2000, for example, it exhibited the GEMÜ 650 Biostar, a diaphragm valve machined out of a solid block of stainless steel and featuring a stainless steel actuator for sterile applications. This was followed up in the same year by the GEMÜ 250 JetStream valve, which was designed according to the bypass principle and has excellent flow properties, making it ideal for use in conjunction with clean to slightly contaminated fluids in the fields of industrial processing and architectural engineering. In 2001, GEMÜ was presented with the Innovation Award of the state of Baden-Württemberg (also known as the Dr. Rudolf Eberle Prize) for its CleanStar range. The metal-free, ultra-pure PFA diaphragm valves in this range are used primarily in the semiconductor and microelectronics industries.

Valves and fittings from the GEMÜ CleanStar range

GEMÜ also identified the potential of the solar energy and photo voltaic market at a very early stage. Although there is no such thing as a 'solar valve',

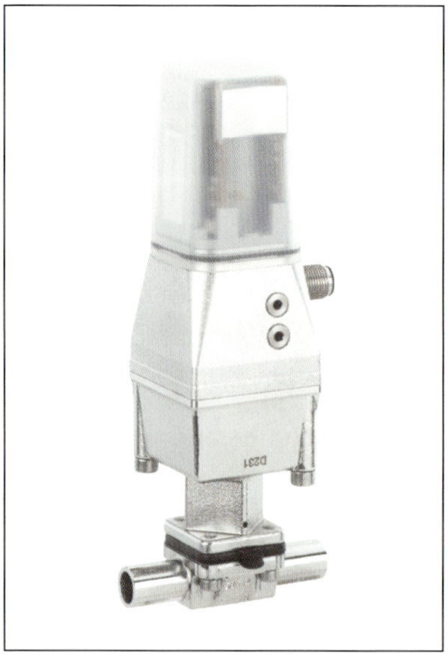

The GEMÜ 651 with integral automation module

the manufacturers of solar panels use a wide selection of valves, pipeline components, and measurement and control systems in their products. Leading manufacturers rely on GEMÜ's broad range and its High Purity products. What began in 2001 with a few isolated orders has since developed into an important area of activity for the company. In view of the fact that business in this sector is booming, GEMÜ expects this unit to continue growing in the coming years. Other new valve types, such as the GEMÜ 550 stainless steel angle seat globe valve for controlling steam and regulating fluids and the GEMÜ 551 KEG (2006), which was developed for the drinks industry, soon followed.

GEMÜ's range of products is continually being extended. The company's aim in doing so is to offer its customers every aspect of valve-related system components. The measurement and control systems sector, for example, has been extended in a targeted manner by the addition of intelligent, 'Bus'-compatible positioners and position indicators as well as level transmitters and level controllers from the LevelLine range. These compact measurement and control components can be combined with elements manufactured by other companies and other modules from GEMÜ's modular system. For example, the GEMÜ 546 valve, which was launched in 2004, builds on a multi-port valve which, when combined with the flat piston actuator designed by Fritz Müller and the GEMÜ 1234 position indicator, creates a very small battery-design valve. The next logical step on from this innovation – i.e. the integration of the actuator and control function in a single component – was realised by GEMÜ in 2006 with the development of the GEMÜ 651 diaphragm valve. The integral automation module in this valve brings together a position indicator, pilot valve, and field bus connector in one single component.

The GEMÜ team at ACHEMA 2006

Having co-operated on a number of occasions with the Esslingen-based automation specialist Festo – one of the world's leading suppliers of pneumatic and electric actuator technology – in the field of high-tech valves and fittings, the two companies entered into a formal co-operation in 2005. Since then, GEMÜ and Festo have been developing products together. Both companies are working together to close the gap between pipe-fitting, plant control, and process control engineering so as to be able to offer plant manufacturers, designers, and operators complete packages that have been tailored to meet their individual application needs. In this regard, they are focussing in particular on the biotechnology and pharmaceutical industries.

GEMÜ's growth in these sectors has been astonishing. Still under Jochen Hofmann's leadership, GEMÜ Sterile Technology (*Steriltechnik*), an independent business unit within the company, currently employs approximately 40 people. The multi-port valves, which are now produced in series, are continually being refined. That being said, the business unit's main focus is on the production of non-standard products and the provision of services related to the design of plants. Ever since GEMÜ Sterile Technology successfully managed the planning and co-ordination of valves and fittings for a major project for an American pharmaceutical company, customers frequently invite the business unit to advise and co-ordinate at a very early stage of the planning process. 'Working hand in hand with the plant manufacturers and the end customer, we develop customised valve and fitting solutions that have been adapted to meet the exact needs of the planned processes,' explains Jochen Hofmann. When it comes to the manufacture of these customised solutions (between 1 and 10 products per batch), GEMÜ Sterile Technology benefits from ultra-modern machinery. The fact that the design and production departments are so closely interlinked makes us incredibly fast and flexible,' explains Hofmann.

Flexibility and speed are not the reserve of GEMÜ Sterile

Tools developed by GEMÜ Precision Technology for non-standard plastic valves

Technology alone: in 2002, for example, GEMÜ's electronics department and the South African sales company OMSA Valves and Instrumentation Ltd. developed a completely new Control Net/AS interface solution for a South African platinum refinery within the space of only eight months. Three years later, GEMÜ's Precision Technology business unit developed injection-moulded plastic valves with a new kind of actuator for a German pharmaceutical company. Only 30 weeks separated the placing of the order and the acceptance of the prototype. This extremely fast implementation was made possible by the combined use of ultra-modern simulation software, 3D rapid prototyping, and 5-axis simultaneous machining.

The speed of its product development and its decades of experience with plastics aided GEMÜ's successful diversification into the rapidly expanding medical market. In 2003, GEMÜ's Swiss subsidiary developed a highly sterile plastic component for the cultivation of transplantation tissue in only four months. This was followed up with more orders. Working closely with doctors and scientists, the Swiss subsidiary has transformed itself into a service centre and centre of excellence for plastics in the medical sector within the space of a few years. All production in Rotkreuz – from the supply of materials through injection moulding to assembly and fully automated packaging – takes place in clean rooms, thereby meeting the incredibly high sterility and product purity standards of the medical sector.

GEMÜ Switzerland's production facility for medical products in Rotkreuz meets the customers' high requirements regarding sterility and product purity

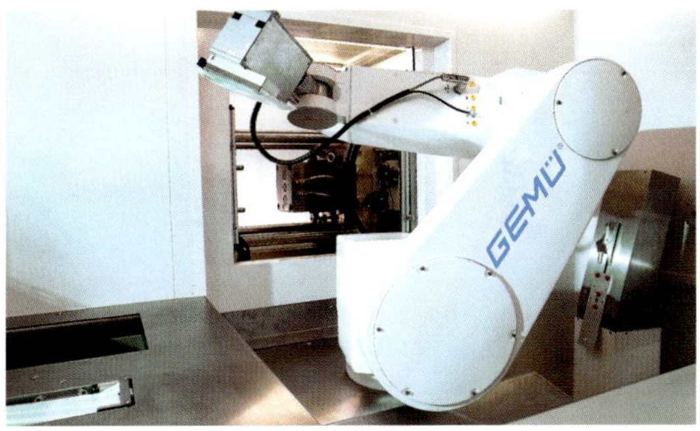

GEMÜ's growth is reflected not only it is turnover, but also in the size of its workforce. In 2008, the GEMÜ group,

which counts a total 18 subsidiaries, employs over 1,000 people. Group headquarters in Criesbach alone, where over 20 new jobs have been created every year since the year 2000, employs approximately 600 people. In order to keep pace with this rapid growth, another extension was necessary only a few years after the 1997/98 extension had been completed. And so, in April 2004, the foundation stone for yet another extension (this time measuring 4,700 square metres) was laid. Half of this area was to be reserved for production. The new building, which cost over 10 million euro and included an automated warehouse for long goods, was opened at a ceremony in the 20-metre high tower in late 2004/ early 2005.

This ceremony coincided with Fritz Müller's 65th birthday and the 40th anniversary of GEMÜ, which was celebrated in January 2004 with friends, business partners, and all members of staff in the City Hall in Neuenstein. The eight-hour celebrations, which included dancing and comedy performances on the theme 'Around the world with GEMÜ' was planned by Gert Müller, his sister Regina, and members of staff.

For Fritz Müller, the jubilee was more than just an opportunity to reflect on the past. Says Müller: 'The older you get, the more awards and honours you receive and the more your company becomes a target for other people's greed. For years, I have been receiving offers of purchase from investors or other companies.' Each offer is rejected using the same standard reply letter. As far as he is concerned, these companies and investors cannot secure GEMÜ's future. 'Takeovers or mergers always put the brakes on a company's ability to innovate, at least for a certain length of time,' he explains. 'But GEMÜ lives off its ability to innovate and off the fact that its staff can identify with their company. You don't need to have the surname "Müller" for that; what is important is that you think and act in the interests of the company. I have taught this maxim to my children, I have used it to hire my staff, and as far as I am concerned, it is the only way to ensure the success of GEMÜ in the long term.'

In 2002, Fritz Müller was presented with the Bundesverdienstkreuz am Bande (Order of Merit of the Federal Republic of Germany) for his services to the region of Hohenlohe and his international social work

Ingrid and Fritz Müller with their children: Regina Kalis (née Müller) and her company VIP-Präsente are responsible for GEMÜ's promotional work, Gert Müller and his company GFM Solutions developed a 3-D prototyping system for GEMÜ and work closely with GEMÜ's Swiss subsidiary in Rotkreuz in the field of medical technology

On his 60th birthday in 1999, Fritz Müller had made it known in public that he was considering gradually withdrawing from the company over the course of the coming five years. On the occasion of his 65th birthday, however, he admitted that this was one of the few ideas he had failed to implement in the course of his working life. One of the reasons for this may be that for those who run family businesses, there is a very thin line between work and private life. 'I founded GEMÜ in the kitchen with the financial support of my mother,' says Fritz Müller. 'My brother and I gave the company its name; my wife was my first employee. Both of my brother's sons work as managers at GEMÜ, my daughter's company is responsible, among other things, for GEMÜ's promotional work, and my son and his company provide us with 3-D prototyping and work closely with our Swiss subsidiary in Rotkreuz in the field of medical products. Most of my early employees were personal friends and have remained loyal to the company even in their retirement. There is hardly any fluctuation in the workforce; on the contrary, sometimes two generations of the same family work here. In other words, GEMÜ itself is a kind of large family.'

Seeing as the company was such a big family, Fritz Müller was all the more devastated to receive word on 25 April 2006 that Herbert Neubauer, his fellow pilot and the man who had built up GEMÜ's Brazilian subsidiary in Curitiba, had been shot during a burglary of his home and was now in hospital with life-threatening wounds. Fritz Müller set off for Curitiba on the very same day. His friend was in a coma and could not be moved. Müller set all possible wheels in motion to ensure

that his friend received the best possible medical treatment and the perpetrators were found. 'I paid the doctors and when I saw that the police had done nothing to follow up the crime, I hired a private detective. I did what I could. They were all quite happy to take the money, but nothing happened. Corruption rules down there!' says Müller, the shock still written in his face. His friend died a few weeks later without ever regaining consciousness. Fritz Müller lost 20 kg weight in the months that followed. 'I was thoroughly shaken by what happened in Brazil; I got sick as a result. I had sent this man to that country for GEMÜ,' he says looking back.

It is possible that this harrowing experience pushed Fritz Müller to realise an idea that he had been harbouring for several years. 'The bigger and more successful GEMÜ became, the more frequently I thought how nice it would be to have a kind of "playground" in addition to our very application-oriented development department; a separate area in which a team of highly qualified engineers could be creative without the restraints of specific work obligations and could develop new ideas; a place where I too, if I felt the need, could go if I wanted to get away from things,' explains the entrepreneur. 'To be honest, a think-tank like this is my idea of happiness. At some point it became clear to me that if I didn't do something about this idea now, nothing would ever come of it.'

This is why Fritz Müller began implementing his 'think-tank' idea in mid 2006. The original idea was that an additional team of developers would be given a set of rooms on the premises in Criesbach. Then, however, a plot of land on a new industrial estate in Waldzimmern, six kilometres from Criesbach, became available. GEMÜ purchased the site, which allowed for the construction of a larger building than would have been feasible in Criesbach. And so, the idea of the GEMÜ Dome, where the new team of engineers would conduct fundamental research separately from the daily business of development in Criesbach, was born. Originally, a museum for Fritz Müller's growing collection of motorcycles and vintage cars was planned to underpin the technical atmosphere of the new

Ingrid and Fritz Müller presenting honours to long-serving GEMÜ employees

building. The plans for a museum were later scrapped, however, when it became clear that GEMÜ would soon once again outgrow the premises in Criesbach and that there was simply not enough space at the site for another extension. Müller decided to house the development department, the testing department with its numerous test benches and testing equipment, and the unit responsible for prototypes and trade fair stands together in the planned GEMÜ Dome in Waldzimmern and to use the entire building as GEMÜ's very own Innovation Centre.

The foundation stone for the GEMÜ Dome was laid in November 2007. The semicircular building designed by the Max Vogelsang company, a modern structure made of reinforced concrete, wood, and glass, will provide 6,000 square metres of space for approximately 150 employees. The intention is that by bringing these departments closer together in the

GEMÜ
Dome

Plans for the GEMÜ Dome

one place and providing them with ultra-modern technology, the time required to bring an idea for a new GEMÜ valve or fitting from development to prototype could be reduced from 1-2 years to only eight weeks.

The heart of this sophisticated, 18-million euro building will be the almost 20-metre high tower, at the top of which is a photovoltaic installation that rotates to catch the maximum amount of sunlight possible. Upon completion, the Dome will house Fritz Müller's office and the engineers' think-tanks, offices, and laboratories.

By the summer of 2008, work on the GEMÜ Dome was progressing well. The anchorage for the glass facades have already been put in place. This will not only give this huge building its transparency and light appearance, but also afford those working inside a wonderful view of the surrounding forest and small lake nearby. 'The intention is that while the atmosphere inside the GEMÜ Dome will be technical, there will always be a link to nature,' says Müller. Then he turns and points to a piece of land near the new building where a tree nursery once stood. 'I worked there in the summer of 1953 before starting my apprenticeship as a precision mechanic. Now, here I am again, starting over from scratch,' he says, laughing.

'AN IDEA IS ONLY GOOD IF YOU TURN IT INTO REALITY'

Fritz Müller, late summer 2008

Herr Müller, you commissioned this book in the hope that your story would encourage other people to set themselves up in business. What does self-employment mean to you?

FM: For me, self-employment means two things: firstly independence, and secondly, the opportunity to turn my ideas into reality. I suppose that has something to do with my family history. My paternal grandfather was, if you like, a self-made man. By the time I was born, he had already built up a thriving business. I grew up with his stories about how he started out with little more than a wheelbarrow, a bucket, a few trowels, and a large dose of ambition. My grandfather's success showed me that it is possible to build up something out of virtually nothing; that it is possible to realise your ambitions as long as you are prepared to work for them. Of course, I also grew up with the story of my father, who ran away from home because he wanted more than anything else to train as a stucco plasterer even though grandfather was against it. And then there was my mother, who after my father's death preferred to give up her comfortable life in Stuttgart and work on the production line in the canning factory in Ingelfingen rather than be dependent on her father-in-law. I obviously got this yearning for independence from both parents. The combination of this striving for independence and the desire to realise my ideas – in other words, my definition of self-employment – has been the most important driving force in my life.

What advice would you give someone who is considering setting up their own business?

FM: Well, as you know, it is always very difficult to give advice. I could make it easy for myself and say: 'You need a good idea at the right moment in time and you need to turn it into a product that can be sold on the market. And if you want to be successful in the long run, you have to keep on searching for new markets.' But of course there is more to it than that. There are lots of other factors that contribute to a person's success as an entrepreneur, but they differ from person to person because everyone's different.'

What factors do you mean?

FM: 'When I look back on my own personal experience, I would say that it helps if you at least like doing whatever it is you will be doing in your own business. Ideally speaking, you should be passionate about it. I was lucky; even as a small boy, I knew without a shadow of a doubt that I wanted to do something with technology and design, simply because these were the things that always interested me. I am still as enthusiastic about my work today as I was then, and I am happy that this is so, because self-employment is about much more than self-realization; it is also about giving 100 per cent. I could not have given 100 per cent over such a long period if I didn't love what I was doing. Quite apart from the fact that it is much easier to interest other people in something – whether it be employees, business partners, or customers – if you really believe in it yourself.

Another point is that you should never be satisfied with what you have achieved at too early a stage; you must make the most of your potential After my apprenticeship, I could have just stayed where I was at Bürkert. My boss was happy with me, everything was going wonderfully well, I was earning good money and could in all probability have worked my way up the company hierarchy. But I didn't feel that I was being challenged enough. I knew that I could do more. So I began studying at the School of Precision Engineering in Schwenningen and I have never once regretted that decision.

Of course, you do need some degree of courage. My mother threw up her hands in horror when I told her that I had given up my good job as head of the testing department at Herion to set myself up in business. I certainly could have stayed at Herion, but when I saw how reluctant they were to pick up on new ideas and how unwilling they were to take risks, I knew that even as the head of the development department at Herion, I would always meet with resistance and objections. But I wanted to be able to implement my ideas without such restrictions. This meant that the path to self-employment was practically already mapped out in front of me.

Naturally, this brings us to the issue of the 'right idea at the right time and in the right form.' Who knows what would have become of me if I had not had the idea for the plastic valve or if I had not succeeded in turning this idea into a marketable product. That was in the days when ideas could be realised with a lathe and a couple of tools; certainly much easier than today where there is hardly any point even making a start unless you have CAD and CNC machines. That being said, for a few years now, there have been such interesting developments, especially in the field of 3-D prototyping, that as far as I am concerned, the entire development and design sector is going through a period of radical change. Another point I would like to make is that it isn't absolutely essential to set yourself up in business in the technical sector. I can imagine that today – just like in my time or my grandfather's time – there are enough opportunities for people to set themselves up in business with a few, simple things. You just have to find your niche and people who will stand by you.

That seems to be a very important aspect. Although you would appear to be the driving force behind GEMÜ, you were always surrounded by people who supported you. How important for your business was the support you received from others?

FM: It was very important indeed. Without this support, things would have been much more difficult for me. Without my grandfather, who always encouraged my desire to set myself up in business; without my mother who always stood surety for me and vouched for me with her good name; without my wife, who took care of the commercial side of things over the course of several years, leaving me free to deal with other matters, and stands by me to this day in everything I do; without my friends, some of whom gave up secure jobs to come and work for me ... without all these people, things would have looked very different. I am very grateful for their support and I believe that as a self-employed person, you need a network, otherwise you won't last very long.

It goes without saying that I also experienced personal disappointment. As the boss, I cannot keep an eye on absolutely everything and have to trust my employees to act in the interests of the company. Unfortunately, there were a number of people over the years who abused this trust. Disappointments like that are very painful. But thankfully, they were the exception rather than the rule. And when I look back at the support my family got from the staff during the worst of GEMÜ's crises – after the fire in 1981 and the flood twelve years later – they more than make up for the disappointment caused by a few black sheep.

Entrepreneurs have to cope not only with themselves and other people, they also have to have a firm grip on finances ...

FM: Yes, if you don't get your finances right, you won't remain an entrepreneur for long and your business will go down the drain. I was always rather careful when it came to financial matters – at least in the early years as an entrepreneur – and I had to learn not to be so timid when it came to money. When we set up a GEMÜ subsidiary in Switzerland, we were offered a site for 200 euro per square metre. I would have had to take out another loan to finance the site and because I considered that too great a risk, I turned down the offer and rented a building instead. Later on when we decided to go ahead and build our own premises, we had to pay 300 euro per square metre for a site that wasn't as good as the first one. Since then, I have adopted a more generous approach to money. Whenever GEMÜ establishes a subsidiary abroad, for example, we need lots of start-up capital to get things moving; it almost always takes several years to get a return on this investment. In such cases, I always say to myself: 'A young tree has to be watered before it can bear fruit', which always makes me sleep better. Of course I am not saying that you should take out loans willy-nilly, but sometimes borrowing capital is absolutely necessary. What's more, a certain amount of pressure from the banks can be a huge motivation.

You mention the word 'pressure'. Some people might look at you and think that you lead a very care-free, self-determined life. After all, you are head of your own company, you jet around the world in your own airplane, and drive vintage cars at the weekend. How do you see your life? Do you personally feel pressure?

FM: When seen from the outside, many people might consider the life I lead to be worth striving for. And to a certain degree it definitely is, when you look at the amount of freedom or the financial opportunities I have. Nevertheless, no-one should be under any illusions: I may fly around the world, but to this day, I have seen very little indeed of the countries I have visited. I am almost always away on business, which means that I see the insides of industrial estates and business parks. At best, I will have enough time to see a tourist attraction or two. Unfortunately, I rarely have time for anything more.

Time is the main problem. I know, for instance, that I did not make enough time for my family, which brings me neatly to the issue of responsibility. Although I can, as an entrepreneur, decide where my company is going, I also bear responsibility for the consequences of my decisions. When it was just me in the business, that was not a problem. However, as soon as I started a family and hired my first member of staff, I assumed a degree of responsibility for these people. I always enjoyed bearing this responsibility, but it certainly wasn't always easy. Thankfully, GEMÜ has always grown and developed. Nevertheless there were always times when I had to make business decisions that could have had serious consequences, not only for my family and I, but also for my employees and their loved ones. At times like that, I still to this day feel huge pressure inside.

How do you deal with that pressure?

FM: What helped me was that there were other things outside work that interested me. These things include my old cars and motorcycles, which interest me as a technician, but are still a kind of toy for me. Above all, flying is very important to me because flying is not a sport for lone warriors. Whenever I am in a jet or an airplane, I almost always have a co-pilot. Even if we don't say much to each other during the flight, we have to rely on each other completely. I value this silent trust greatly. With gliding, things are different; you are generally up there alone. But once you land, it is all about camaraderie and the club. Even though my wife and I could have afforded to go away on luxury holidays, we preferred to spend our summer holidays in Austria and our winter holidays skiing with our friends from the club. Although I very rarely get the chance to glide these days, being actively involved in the club and enjoying the contact with the other pilots is still very important to me. For them, I'm still Fritz. This kind of contact is very important for me; otherwise you run the risk of forgetting where you come from. It's only when you know where you come from that you can decide where you want to go. Knowing where you come from and where you want to go is one the most important prerequisites for independence, regardless of whether you actually set yourself up in business or whether you are quite simply someone who thinks and acts independently.

Let me ask you about renewal and innovation. Throughout our co-operation on this book, you repeatedly stressed the importance of innovation. Why do you consider innovation to be so important?

FM: The reason is directly linked to what I just said. Nowadays, whenever I talk to people in Asia, Brazil, or Russia, I always notice that they know exactly where they want to go. They want a better life, and when you look at what daily life is like for most people in these countries, this desire is perfectly understandable. This hunger for a better life triggers a similar ability to renew and innovate to the one that people here in Germany developed in the years directly after the war. If we in Germany want to maintain the lead we worked so hard to achieve back then, we cannot stand still; we have to keep on moving, keep on looking for new paths, keep on developing new things. If we don't grasp this fact, we will soon be left behind. Only those who are capable of constant renewal have a future. That is why innovation is so important.

What are the prerequisites for innovation?

FM: I think that it is a mistake to believe that innovations come about spontaneously and without any form of preparation. Of course, there will always be those moments when you get a brainwave; all of a sudden the road ahead is clear. However, in the majority of cases, you only get that bright idea after dealing with a specific problem for months, if not years. When you get that idea, you've got to turn it into reality. That costs a lot of money, especially when you have a team of designers and developers working on the project. Being open to new ideas, a readiness to invest, and a good deal of staying power are, therefore, extremely important prerequisites for innovation. The best example is the development of GEMÜ's aseptic systems business. In the early 1990s, our High Purity unit was developing faster than any other in the company. At that time, we only had standard products in our biotechnology range, and while our turnover in this sector was stable, it was certainly not very high. If I had been out to make a quick buck, then I would have developed the HP unit and closed down the biotechnology unit. Instead, I gave Jochen Hofmann a dedicated budget and told him to take a closer look at the biotechnology sector. The first thing Hofmann did was to develop a completely new approach to customer work and product management. We were soon reaping the benefits of this new approach. The real breakthrough came five years later with the launch of multi-port valves. Even that was only possible because we had previously invested in the modernisation of our machinery. In other words, we had to invest heavily over the course of a few years in order to create both the mental and technical conditions that allowed us to develop this innovative product. We could afford to do so because GEMÜ was not looking for short-term results, but was focussing instead on long-term success. Today, despite the continuing crisis in the semiconductor industry, we are still doing good business in the HP sector, but we are now generating almost half of our turnover with products for the Bio-pharmaceutical industry.

How important is education for you in all this?

FM: It plays a key role in all this and that is why we are so involved in education. The best ideas and the most modern machinery are of no use whatsoever if you don't have the people to implement the ideas and to operate the machines. Moreover, if I want to sell my products all over the world, I need people who speak foreign languages, can move with ease in different cultures, and are maybe even willing to spend some time building up a subsidiary in a different country. Then there is, of course, the quality aspect. To this day, products that are made in Germany are respected throughout the world, not least because of their high quality. If we cannot at least maintain this standard, we will not be able to withstand the competition from emerging economies in the medium term. However, in order to maintain high quality standards, we need highly qualified employees, not only at management level, but in all areas of the company. That too costs money. Nevertheless, I am convinced that this is the most worthwhile investment of all. The final prerequisite for innovation – and as far as I am concerned, the most important one of all – doesn't cost a thing: hard work, hard work, and more hard work. Although this virtue may now be considered by some in Germany to be somewhat outdated, we in the Hohenlohe region are lucky enough to be a little behind the times! When I look at my staff in Criesbach, I am confident that we have what it takes to remain innovative and therefore competitive for years to come. This is why GEMÜ has faith in Germany as an industrial location.

ACKNOWLEDGEMENTS

The author, Dr Susanne Saygin, and the editorial team at Geschichtsbüro Reder, Roeseling & Prüfer would like to thank the following people for their valuable support in the making of this book. This project could not have been completed without their assistance.

First and foremost, we should like to thank Fritz Müller. His openness in the interviews we conducted with him made it possible to produce the text in its present form.

Ingrid Müller was kind enough to grant us a number of interviews. She also made a large number of photographs available and showed great interest in the project.

Manfred Schmitt, Head of Ingelfingen Town Hall, granted us access to Ingelfingen's archives and provided us with images and photographs relating to events in local history.

Professor Dr Axel Schenk of the Reinhold Würth School (RWH) in Künzelsau gave us an overview of the courses on offer at the school, which is supported by GEMÜ.

Günter Nägele and Helmut Weber, school friends of Fritz Müller and two of his earliest employees, provided important information about life in Ingelfingen after the war and the early years of GEMÜ. Emil Fick and Manfred Hofmann, long-serving GEMÜ employees, completed the picture with their stories and personal memories.

We are also grateful to the following employees for giving us not only guided tours, but also an insight into their respective departments: Siegfried Hoffmann (Commercial Managing Director), Jochen Hofmann (Head of Sterile Technology), Ilka Rölke (Human Resources), Jörg Röttgen (Head of Commercial Department), Alois Walter (Head of Training (Electronic Engineering)), Hermann Walter (Works Manager), and Wolfgang Wick (Head of Training (Metalwork and Mechanical Engineering)).

Ralph Kroupa and Manfred Zeiser were on hand to answer our technical questions about specific aspects of valve and fitting technology.

Dorothea Mayer, Fritz Müller's Personal Assistant, made sure that communication with Fritz Müller always functioned quickly and smoothly despite his many professional engagements and foreign trips. We are particularly indebted to her and to Teresa Bareis, who was responsible for co-ordinating the book project and for clarifying countless organisational matters.

SELECTED BIBLIOGRAPHY

The primary source of information for this biography was a series of eight interviews with Fritz Müller himself, each of which lasted several hours and was conducted by Dr Susanne Saygin and Dr Dirk Reder of Geschichtsbüro Reder, Roeseling & Prüfer between September 2007 and June 2008. Lengthy interviews were also conducted with Ingrid Müller, Jochen Hofmann, Günter Nägele, Hermann Walter, and Helmut Weber. These interviews provided the general outline for the story. Any gaps in the biography were filled in by interviews with Ilka Rölke, Emil Fick, Manfred Hofmann, Jörg Röttgen, Alois Walter, and Wolfgang Wick as well as the brief history of GEMÜ written by GEMÜ employee Volker Wagner in 2004.

The Müller family kindly made personal documents available to the author for her reconstruction of the history of Fritz Müller's family life. The lion's share of the images and photos in this book were taken from the family's private photograph albums. A variety of archives in the state of Baden-Württemberg also made pictures and images available.

The GEMÜ company archive was almost completely destroyed by the fire that ravaged the company premises in 1982 and the serious floods of 1993. For this reason, press reports on GEMÜ and portraits of Fritz Müller that were printed in the regional press and in particular the *Hohenloher Zeitung* and the *Heilbronner Stimme* were used as sources of information. Another important source was the complete set of *GEMÜ-News*, a newspaper for customers and employees which has been published by GEMÜ since 1999.

The description of some of GEMÜ's technical developments are based on a collection of articles on GEMÜ products published in specialist magazines compiled by Ralph Kroupa at GEMÜ.

Specialist literature on valves and fittings and on the history of the Hohenlohe region were also used. The most important sources of information, newspaper and magazine articles, and key works are listed in the selected bibliography below.

Interviews

Interviews with Fritz Müller on 21/22.09.2007, 11.12.2007, 29/30.01.2008, 26/27.02.2008, and 19.06.2008.

Interview with Ingrid Müller on 23.04.2008.

Interview with Hermann Walter on 23.04.2008.

Interview with Günter Nägele and Helmut Weber on 23.04.2008

Interview with Jochen Hofmann on 25.04.2008.

Other information was gleaned in interviews with Emil Fick, Manfred Hofmann, Ralph Kroupa, Ilka Rölke, Alois Walter, and Wolfgang Wick.

Manuscripts

'10 Jahre GEMÜ', speech made on the occasion of the tenth anniversary of the establishment of GEMÜ, manuscript kept in a photo album belonging to the Müller family.

'1892-1992 – 100 Jahre Weingärtnergenossenschaft Ingelfingen', now: Kochertalkellerei eG, speech made by Heinrich Ehrmann, retired mayor, Ingelfingen, manuscript.

'Abschied für den Soldaten Kristian Hasel, von Ingelfingen', original document recording the honourable discharge of Fritz Müller's ancestor, Kristian Hasel, from the army, dated 25.03.1820 and now in the possession of the Müller family.

Report about research into the Müller family crest, taken from Kalckhoff, A., Heraldry International, 1988.

'Extrablatt', a newspaper made by Ingrid Müller on the occasion of her husband's 60th birthday in 1998.

'Funkbremsbares Bobby-Car', GEMÜ entry for the German national 'Young People Research' competition in 2007 (presentation folder).

Ruchser, E., Technik in meinem Leben, published by the author, no year, no location given.

Company overview of the apprenticeship and continuing professional development opportunities at GEMÜ compiled by Ilka Rölke, January 2008.

Wagner, V., 'Die Geschichte von GEMÜ', handwritten history of GEMÜ, now in the possession of the Müller family, 2004.

Selected articles from GEMÜ-News

'Bundesverdienstkreuz für Fritz Müller', *GEMÜ-News*, 05/2000, 1.

'GEMÜ erweitert Standort Ingelfingen', *GEMÜ-News* 02/2004, 1.

'GEMÜ Schweiz feiert ihr 25jähriges Jubiläum', *GEMÜ-News*, 03/2006, 3.

Hofmann, J., 'Zukunftsmarkt Steriltechnik', *GEMÜ-News* 02/2002, 1.

Plastic solutions for medical devices (Hauspublikation der GEMÜ Schweiz), 01/2005.

Schmezer, U., 'Keine Chance für Morgenmuffel', *GEMÜ-News*, 01/2007, 3.

Schulz, T., 'Automatisieren ohne Druckluft', *GEMÜ-News*, 01/2006, 1.

'Sonderprojekt im Kunststoffspritzguss', *GEMÜ-News*, 02/2005, 3.

'Wirtschaftswunder Hohenlohe', *GEMÜ-News*, 02/2004, 6.

Zhen Xiao, 'GEMÜ China – jung, dynamisch und erfolgreich', *GEMÜ-News*, 02/2002, 2.

Books, newspaper articles, and articles in specialist magazines

'35 Jahre InterApp', *cav*, 05/2005, 35.

50 Jahre bürkert, Ingelfingen, 1996.

'Armaturen und Ventile für moderne Prozessleitsysteme', *3R international*, 05/2006, 239.

'Aseptische Membranventile', *CHEManager* 09/2000, 25.

'Aus einem Stück', *chemie-anlagen+verfahren*, 04/2004, 58.

'Automation inside', *SPS+Prozess*, 09/2006, 126-127.

Bauschert, O. (Hrsg.), Hohenlohe, Stuttgart, Berlin, Köln, 1993.

Beschreibung des Oberamts Künzelsau, Stuttgart, 1883.

Brinkmann, G./Zacher, H., Die Evolution der Segelflugzeuge, Bonn, 1992.

'Eine Idee ist nur gut, wenn man etwas daraus macht', *Hohenloher Zeitung*, 02.01.2004, (o.S.).

'Ein offenes Ohr für jedermann', *Hohenloher Zeitung*, 02.01.2004 (o.S.).

Fischer, W., Das Fürstentum Hohenlohe im Zeitalter der Aufklärung, Tübingen, 1958.

'Für alle Auszubildenden auch ein Arbeitsplatz', *Hohenloher Zeitung*, 10.08.2000 (o.S.).

Geistmann, D., Segelflugzeuge in Deutschland, Stuttgart, 1992.

Gräter, D., Die Kochertalbahn. Waldenburg – Künzelsau – Forchtenberg, Künzelsau, 2003.

Griesinger, B., 'Schmiede für High-Tech Ideen', *Hohenloher Zeitung*, 09.11.2007, (o.S.).

Hämmerle, S., 'Firmen- und Gemeinwohl zugleich gefördert', *Hohenloher Zeitung*, 11.06.2005, (o.S.).

'Heimatverwurzelt und weltoffen', *Hohenloher Zeitung*, 05.01.2004 (o.S.).

Der Hohenlohekreis, 2 Bde., Ostfildern, 2006.

'Ideen en masse und eine Belegschaft, die zusammenhält', Wirtschaftstimme (Beilage der *Heilbronner Stimme*), 27.05.2003, 27.

'Intelligente Armaturen und Ventile', Prozesstechnik und Automation, März 2006, o.S.

Kleemann, G., Schmitt, C.L., *Das Hohenloher Land*, Blaufelden, 1993.

Kroupa. R., *Ventiltechnologie im Anlagenbau* (Weinheim, 1994).

Lindauer, G., *Beiträge zur Verstädterung in ländlichen Räumen. Mit Beispielen aus dem Kochertal*, Stuttgart, 1971.

Meier, R., *Hohenlohe in alten Zeiten*, Künzelsau, 2004.

Messer, R., *Es war einmal in Oberursel. Vom Kinderautomobil zur Seifenkiste*, Erfurt, 2007.

Müller, F./Hagdorn, H., *Das Ingelfinger Fass. Weinbaugeschichte im Kochertal*, Ingelfingen, 2002.

Rauser, J.H., *Ingelfinger Heimatbuch*, Ingelfingen, 1980.

Robisheaux, T., *Rural society and the search for order in early modern Germany*, Cambridge, New York u.a., 1989.

Schenk, H. K., *Hohenlohe- vom Reichsfürstentum zur Standesherrschaft*, Künzelsau, 2006.

Schlick, G., *Ventile in unserem Leben*, Würzburg, 1978.

Schmidt, W., 'Bodenständig und weltoffen zugleich', *Hohenloher Zeitung*, 14.04.2000, (o.S.).

Schremmer, E., *Die Bauernbefreiung in Hohenlohe*, Stuttgart, 1963.

Schwarz, M., 'Waldzimmern wird Dome-Stadt', *Hohenloher Zeitung*, 1.8.2007, (o.S.).

Steinle, P., *Die Vermögensverhältnisse der Landbevölkerung in Hohenlohe im 17. und 18. Jahrhundert*, Stuttgart, 1971.

'Sterile Mehrwegeventile', *Steriltechnik*, 01/2004, 116.

Stockburger, M., 'Festo kooperiert mit Ingelfinger GEMÜ', *Heilbronner Stimme*,19.10.2005, (o.S.).

Stockburger, M., 'Kathedrale der Technologie', *Hohenloher Zeitung*, 17.06.2008, (o.S.)

Theiss, K./Their, M. (Hrsg.), *Der Kreis Künzelsau*, Aalen, Stuttgart, 1965.

The U.S. Armed Forces German Youth Activities Program 1945-1955, Historical Division Headquarters, United States Army, Europe, 1956.

'Den Ventilanwendern das Optimum anbieten ...', *Verfahrenstechnik* 29 (1995), Nr. 10, 44-47.

'Viel Funktion auf engstem Raum', *Metalloberfläche*, 12/2004, 12.

Weber, W., *Das Bastelbuch für Vater und Sohn*, Stuttgart, 11954, 21958.

Weik, H., *Die Agrar- und Wirtschaftsverhältnisse des Fürstentums Hohenlohe im 18. Jahrhundert*, Köln, 1969.

'Zuerst die Firma und dann der Chef', *Hohenloher Zeitung*, 02.01.2004 (o.S.).

PICTURE CREDITS

p. 15, p. 24, p. 29, p. 63: Müller, F. and Hagdorn, H., *Das Ingelfinger Fass. Weinbaugeschichte im Kochertal*, Ingelfingen, 2002 (taken from p. 45, p. 43, p. 25, p. 44, p. 27).

p. 19, p. 21: *Der Hohenlohekreis*. 2 Bde. Ostfildern, 2006, Vol. 1, p. 38 and Vol. 2, map insert (all rights reserved by the Landesarchiv Baden-Württemberg).

p. 22: Landesamt für Denkmalpflege, Esslingen.

p. 23: Regierungspräsidium Karlsruhe, Referat 25 Denkmalpflege.

p. 25, p. 33, p. 58: Rauser, J. H., *Ingelfinger Heimatbuch*, Erstes Buch Stadtgeschichte, Ingelfingen, 1980 (taken from p. 63, p. 261, p. 253).

p. 28, p. 55, p. 56, p. 57: Stadtarchiv Ingelfingen.

p. 40: Kreisarchiv Hohenlohe, Neuenstein.

p. 43, p. 47, p. 48: akg images.

p. 69: Herion Informationen, 2/3 (1962), p. 12.

p. 70, p. 81: *Herion – Schrittmacher der Automation*, Fellbach, 1968 (taken from p. 1, p. 103).

p. 78, p. 79: Schlick, G., *Ventile in unserem Leben*, Würzburg, 1978, p. 196.

p. 83: Geistmann, D., *Segelflugzeuge in Deutschland*, Stuttgart, 1992.

p. 118, p. 119: Kroupa, R., *Ventiltechnologie im Anlagenbau*, Weinheim, 1994 (taken from p. 202, p. 178).

All other illustrations were taken from the Müller family's private collection or from GEMÜ.